Verification & Validation of Selected Fire Models for Nuclear Power Plant Applications

Volume 6: MAGIC

NUREG-1824   **EPRI 1011999**

Final Report

May 2007

U.S. Nuclear Regulatory Commission
Office of Nuclear Regulatory Research (RES)
Two White Flint North, 11545 Rockville Pike
Rockville, MD 20852-2738

U.S. NRC-RES Project Manager
M. H. Salley

Electric Power Research Institute (EPRI)
3420 Hillview Avenue
Palo Alto, CA 94303

EPRI Project Manager
R.P. Kassawara

DISCLAIMER OF WARRANTIES AND LIMITATION OF LIABILITIES

NOTE

CITATIONS

This report was prepared by

U.S. Nuclear Regulatory Commission,
Office of Nuclear Regulatory Research (RES)
Two White Flint North, 11545 Rockville Pike
Rockville, MD 20852-2738

Principal Investigators:
K. Hill
J. Dreisbach

Electric Power Research Institute (EPRI)
3420 Hillview Avenue
Palo Alto, CA 94303

Science Applications International Corp (SAIC)
4920 El Camino Real
Los Altos, CA 94022

Principal Investigators:
F. Joglar
B. Najafi

National Institute of Standards and Technology
Building Fire Research Laboratory (BFRL)
100 Bureau Drive, Stop 8600
Gaithersburg, MD 20899-8600

Principal Investigators:

K McGrattan

R. Peacock

A. Hamins

Volume 1, Main Report: B. Najafi, F. Joglar, J. Dreisbach
Volume 2, Experimental Uncertainty: A. Hamins, K. McGrattan
Volume 3, FDT^S: J. Dreisbach, K. Hill
Volume 4, FIVE-Rev1: F. Joglar
Volume 5, CFAST: R. Peacock, P. Reneke (NIST)
Volume 6, MAGIC: F. Joglar, B. Guatier (EdF), L. Gay (EdF), J. Texeraud (EdF)
Volume 7, FDS: K. McGrattan

This report describes research sponsored jointly by U.S. Nuclear Regulatory Commission, Office of Nuclear Regulatory Research (RES) and Electric Power Research Institute (EPRI).

The report is a corporate document that should be cited in the literature in the following manner:

Verification and Validation of Selected Fire Models for Nuclear Power Plant Applications, Volume 6: MAGIC, U.S. Nuclear Regulatory Commission, Office of Nuclear Regulatory Research (RES), Rockville, MD, 2007, and Electric Power Research Institute (EPRI), Palo Alto, CA, NUREG-1824 and EPRI 1011999.

ABSTRACT

There is a movement to introduce risk-informed and performance-based analyses into fire protection engineering practice, both domestically and worldwide. This movement exists in the general fire protection community, as well as the nuclear power plant (NPP) fire protection community. The U.S. Nuclear Regulatory Commission (NRC) has used risk-informed insights as part of its regulatory decision making since the 1990's.

In 2002, the National Fire Protection Association (NFPA) developed NFPA 805, *Performance-Based Standard for Fire Protection for Light-Water Reactor Electric Generating Plants, 2001 Edition*. In July 2004, the NRC amended its fire protection requirements in Title 10, Section 50.48, of the *Code of Federal Regulations* (10 CFR 50.48) to permit existing reactor licensees to voluntarily adopt fire protection requirements contained in NFPA 805 as an alternative to the existing deterministic fire protection requirements. In addition, the NPP fire protection community has been using risk-informed, performance-based (RI/PB) approaches and insights to support fire protection decision-making in general.

One key tool needed to further the use of RI/PB fire protection is the availability of verified and validated fire models that can reliably predict the consequences of fires. Section 2.4.1.2 of NFPA 805 requires that only fire models acceptable to the Authority Having Jurisdiction (AHJ) shall be used in fire modeling calculations. Furthermore, Sections 2.4.1.2.2 and 2.4.1.2.3 of NFPA 805 state that fire models shall only be applied within the limitations of the given model, and shall be verified and validated.

This report is the first effort to document the verification and validation (V&V) of five fire models that are commonly used in NPP applications. The project was performed in accordance with the guidelines that the American Society for Testing and Materials (ASTM) set forth in ASTM E 1355, *Standard Guide for Evaluating the Predictive Capability of Deterministic Fire Models*. The results of this V&V are reported in the form of ranges of accuracies for the fire model predictions.

FOREWORD

Fire modeling and fire dynamics calculations are used in a number of fire hazards analysis (FHA) studies and documents, including fire risk analysis (FRA) calculations; compliance with and exemptions to the regulatory requirements for fire protection in 10 CFR Part 50; the Significance Determination Process (SDP) used in the inspection program conducted by the U.S. Nuclear Regulatory Commission (NRC); and, most recently, the risk-informed performance-based (RI/PB) voluntary fire protection licensing basis established under 10 CFR 50.48(c). The RI/PB method is based on the National Fire Protection Association (NFPA) Standard 805, *Performance-Based Standard for Fire Protection for Light-Water Reactor Generating Plants*.

The seven volumes of this NUREG-series report provide technical documentation concerning the predictive capabilities of a specific set of fire dynamics calculation tools and fire models for the analysis of fire hazards in postulated nuclear power plant (NPP) scenarios. Under a joint memorandum of understanding (MOU), the NRC Office of Nuclear Regulatory Research (RES) and the Electric Power Research Institute (EPRI) agreed to develop this technical document for NPP application of these fire modeling tools. The objectives of this agreement include creating a library of typical NPP fire scenarios and providing information on the ability of specific fire models to predict the consequences of those typical NPP fire scenarios. To meet these objectives, RES and EPRI initiated this collaborative project to provide an evaluation, in the form of verification and validation (V&V), for a set of five commonly available fire modeling tools.

The road map for this project was derived from NFPA 805 and the American Society for Testing and Materials (ASTM) Standard E 1355, *Standard Guide for Evaluating the Predictive Capability of Deterministic Fire Models*. These industry standards form the methodology and process used to perform this study. Technical review of fire models is also necessary to ensure that those using the models can accurately assess the adequacy of the scientific and technical bases for the models, select models that are appropriate for a desired use, and understand the levels of confidence that can be attributed to the results predicted by the models. This work was performed using state-of-the-art fire dynamics calculation methods/models and the most applicable fire test data. Future improvements in the fire dynamics calculation methods/models and additional fire test data may impact the results presented in the seven volumes of this report.

This document does not constitute regulatory requirements, and NRC participation in this study neither constitutes nor implies regulatory approval of applications based on the analysis contained in this text. The analyses documented in this report represent the combined efforts of individuals from RES and EPRI. Both organizations provided specialists in the use of fire models and other FHA tools to support this work. The results from this combined effort do not constitute either a regulatory position or regulatory guidance. Rather, these results are intended to provide technical analysis of the predictive capabilities of five fire dynamic calculation tools, and they may also help to identify areas where further research and analysis are needed.

Brian W. Sheron, Director
Office of Nuclear Regulatory Research
U.S. Nuclear Regulatory Commission

CONTENTS

FIGURES

TABLES

REPORT SUMMARY

This report documents the verification and validation (V&V) of five selected fire models commonly used in support of risk-informed and performance-based (RI/PB) fire protection at nuclear power plants (NPPs).

Background

Since the 1990s, when it became the policy of the NRC to use risk-informed methods to make regulatory decisions where possible, the nuclear power industry has been moving from prescriptive rules and practices toward the use of risk information to supplement decision-making. Several initiatives have furthered this transition in the area of fire protection. In 2001, the National Fire Protection Association (NFPA) completed the development of NFPA Standard 805, *Performance-Based Standard for Fire Protection for Light-Water Reactor Electric Generating Plants*, 2001 Edition. Effective July 16, 2004, the NRC amended its fire protection requirements in Title 10, Section 50.48(c), of the *Code of Federal Regulations* [10 CFR 50.48(c)] to permit existing reactor licensees to voluntarily adopt fire protection requirements contained in NFPA 805 as an alternative to the existing deterministic fire protection requirements. RI/PB fire protection often relies on fire modeling for determining the consequence of fires. NFPA 805 requires that the "fire models shall be verified and validated," and "only fire models that are acceptable to the Authority Having Jurisdiction (AHJ) shall be used in fire modeling calculations."

Objectives

- To perform V&V studies of selected fire models using a consistent methodology (ASTM I 1335)

- To investigate the specific fire modeling issue of interest to NPP fire protection applications

- To quantify fire model predictive capabilities to the extent that can be supported by comparison with selected and available experimental data.

Approach

This project team performed V&V studies on five selected models: (1) NRC's NUREG-1805 Fire Dynamics Tools (FDTS), (2) EPRI's Fire-Induced Vulnerability Evaluation Revision 1 (FIVE-Rev1), (3) National Institute of Standards and Technology's (NIST) Consolidated Model of Fire Growth and Smoke Transport (CFAST), (4) Electricité de France's (EdF) MAGIC, and (5) NIST's Fire Dynamics Simulator (FDS). The team based these studies on the guidelines of the ASTM E 1355, *Standard Guide for Evaluating the Predictive Capability of Deterministic Fire Models*. The scope of these V&V studies was limited to the capabilities of the selected fire models and did not cover certain potential fire scenarios that fall outside the capabilities of these fire models.

Results

The results of this study are presented in the form of relative differences between fire model predictions and experimental data for fire modeling attributes such as plume temperature that are important to NPP fire modeling applications. While the relative differences sometimes show agreement, they also show both under-prediction and over-prediction in some circumstances. These relative differences are affected by the capabilities of the models, the availability of accurate applicable experimental data, and the experimental uncertainty of these data. The project team used the relative differences, in combination with some engineering judgment as to the appropriateness of the model and the agreement between model and experiment, to produce a graded characterization of each fire model's capability to predict attributes important to NPP fire modeling applications.

This report does not provide relative differences for all known fire scenarios in NPP applications. This incompleteness is attributable to a combination of model capability and lack of relevant experimental data. The first problem can be addressed by improving the fire models, while the second problem calls for more applicable fire experiments.

EPRI Perspective

The use of fire models to support fire protection decision-making requires a good understanding of their limitations and predictive capabilities. While this report makes considerable progress toward this goal, it also points to ranges of accuracies in the predictive capability of these fire models that could limit their use in fire modeling applications. Use of these fire models presents challenges that should be addressed if the fire protection community is to realize the full benefit of fire modeling and performance-based fire protection. Persisting problems require both short-term and long-term solutions. In the short-term, users need to be educated on how the results of this work may affect known applications of fire modeling, perhaps through pilot application of the findings of this report and documentation of the resulting lessons learned. In the long-term, additional work on improving the models and performing additional experiments should be considered.

Keywords

Fire

Verification and Validation (V&V)

Risk-Informed Regulation

Fire Safety

Nuclear Power Plant

Fire Probabilistic Safety Assessment (PSA)

Fire Modeling

Performance-Based

Fire Hazard Analysis (FHA)

Fire Protection

Fire Probabilistic Risk Assessment (PRA)

PREFACE

This report is presented in seven volumes. Volume 1, the Main Report, provides general background information, programmatic and technical overviews, and project insights and conclusions. Volume 2 quantifies the uncertainty of the experiments used in the V&V study of the five fire models considered in this study. Volumes 3 through 7 provide detailed discussions of the verification and validation (V&V) of the fire models:

Volume 3   Fire Dynamics Tools (FDT^s)

Volume 4   Fire-Induced Vulnerability Evaluation, Revision 1 (FIVE-Rev1)

Volume 5   Consolidated Model of Fire Growth and Smoke Transport (CFAST)

Volume 6   MAGIC

Volume 7   Fire Dynamics Simulator (FDS)

ACKNOWLEDGMENTS

The work documented in this report benefited from contributions and considerable technical support from several organizations.

The verification and validation (V&V) studies for FDT^s (Volume 3), CFAST (Volume 5), and FDS (Volume 7) were conducted in collaboration with the U.S. Department of Commerce, National Institute of Standards and Technology (NIST), Building and Fire Research Laboratory (BFRL). Since the inception of this project in 1999, the NRC has collaborated with NIST through an interagency memorandum of understanding (MOU) and conducted research to provide the necessary technical data and tools to support the use of fire models in nuclear power plant fire hazard analysis (FHA).

We appreciate the efforts of Doug Carpenter and Rob Schmidt of Combustion Science Engineers, Inc. for their comments and contributions to Volume 3.

In addition, we acknowledge and appreciate the extensive contributions of Electricité de France (EdF) in preparing Volume 6 for MAGIC.

We thank Drs. Charles Hagwood and Matthew Bundy of NIST for the many helpful discussions regarding Volume 2.

We also appreciate the efforts of organizations participating in the International Collaborative Fire Model Project (ICFMP) to Evaluate Fire Models for Nuclear Power Plant Applications, which provided experimental data, problem specifications, and insights and peer comment for the international fire model benchmarking and validation exercises, and jointly prepared the panel reports used and referred to in this study. We specifically appreciate the efforts of the Building Research Establishment (BRE) and the Nuclear Installations Inspectorate in the United Kingdom, which provided leadership for ICFMP Benchmark Exercise (BE) #2, as well as Gesellschaft für Anlagen-und Reaktorsicherheit (GRS) and Institut für Baustoffe, Massivbau und Brandschutz (iBMB) in Germany, which provided leadership and valuable experimental data for ICFMP BE #4 and BE #5. In particular, ICFMP BE #2 was led by Stewart Miles at BRE; ICFMP BE #4 was led by Walter Klein-Hessling and Marina Rowekamp at GRS, and R. Dobbernack and Olaf Riese at iBMB; and ICFMP BE #5 was led by Olaf Riese and D. Hosser at iBMB, and Marina Rowekamp at GRS. Simo Hostikka of VTT, Finland also assisted with ICFMP BE#2 by providing pictures, tests reports, and answered various technical questions of those experiments. We acknowledge and sincerely appreciate all of their efforts.

We greatly appreciate Paula Garrity, Technical Editor for the Office of Nuclear Regulatory Research, and Linda Stevenson, agency Publications Specialist, for providing editorial and publishing support for this report. Lionel Watkins and Felix Gonzalez developed the graphics for Volume 1. We also greatly appreciate Dariusz Szwarc and Alan Kouchinsky for their assistance finalizing this report.

We wish to acknowledge the team of peer reviewers who reviewed the initial draft of this report and provided valuable comments. The peer reviewers were Dr. Craig Beyler and Mr. Phil DiNenno of Hughes Associates, Inc., and Dr. James Quintiere of the University of Maryland.

Finally, we would like to thank the internal and external stakeholders who took the time to provide comments and suggestions on the initial draft of this report when it was published in the *Federal Register* (71 FR 5088) on January 31, 2006. Those stakeholders who commented are listed and acknowledged below.

Janice Bardi, ASTM International

Moonhak Jee, Korea Electric Power Research Institute

U.S. Nuclear Regulatory Commission, Office of Nuclear Reactor Regulation Fire Protection Branch

J. Greg Sanchez, New York City Transit

David Showalter, Fluent, Inc.

Douglas Carpenter, Combustion Science & Engineering, Inc.

Nathan Siu, U.S. Nuclear Regulatory Commission, Office of Nuclear Regulatory Research

Clarence Worrell, Pacific Gas & Electric

LIST OF ACRONYMS

AGA	American Gas Association
AHJ	Authority Having Jurisdiction
ASME	American Society of Mechanical Engineers
ASTM	American Society for Testing and Materials
BE	Benchmark Exercise
BFRL	Building and Fire Research Laboratory
BRE	Building Research Establishment
BWR	Boiling-Water Reactor
CDF	Core Damage Frequency
CFAST	Consolidated Fire Growth and Smoke Transport Model
CFD	Computational Fluid Dynamics
CFR	*Code of Federal Regulations*
CSR	Cable Spreading Room
EdF	Electricité de France
EPRI	Electric Power Research Institute
FDS	Fire Dynamics Simulator
FDT^s	Fire Dynamics Tools (NUREG-1805)
FHA	Fire Hazard Analysis
FIVE-Rev1	Fire-Induced Vulnerability Evaluation, Revision 1
FMRC	Factory Mutual Research Corporation
FM/SNL	Factory Mutual & Sandia National Laboratories
FPA	Foote, Pagni, and Alvares
FRA	Fire Risk Analysis
GRS	Gesellschaft für Anlagen-und Reaktorsicherheit (Germany)
HGL	Hot Gas Layer

HRR	Heat Release Rate
IAFSS	International Association of Fire Safety Science
iBMB	Institut für Baustoffe, Massivbau und Brandschutz
ICFMP	International Collaborative Fire Model Project
IEEE	Institute of Electrical and Electronics Engineers
IPEEE	Individual Plant Examination of External Events
LOL	Low Oxygen Limit
MCC	Motor Control Center
MCR	Main Control Room
MQH	McCaffrey, Quintiere, and Harkleroad
MOU	Memorandum of Understanding
NBS	National Bureau of Standards (now NIST)
NFPA	National Fire Protection Association
NIST	National Institute of Standards and Technology
NPP	Nuclear Power Plant
NRC	U.S. Nuclear Regulatory Commission
NRR	Office of Nuclear Reactor Regulation (NRC)
ODE	Ordinary Differential Equation
PMMA	Polymethyl-methacrylate
PWR	Pressurized Water Reactor
RCP	Reactor Coolant Pump
RES	Office of Nuclear Regulatory Research (NRC)
RI/PB	Risk-Informed, Performance-Based
SBDG	Stand-By Diesel Generator
SDP	Significance Determination Process
SFPE	Society of Fire Protection Engineers
SNL	Sandia National Laboratories
SWGR	Switchgear Room
V&V	Verification & Validation

1
INTRODUCTION

As the use of fire modeling tools increases in support of day-to-day nuclear power plant applications the importance of verification and validation (V&V) studies for these tools also increases. V&V studies afford fire modeling analysts confidence in applying analytical tools by quantifying and discussing the performance of the given model in predicting the fire conditions measured in a particular experiment. The underlying assumptions, capabilities, and limitations of the model are discussed and evaluated as part of the V&V study.

The main objective of this report is to document a V&V study for the MAGIC zone model, in accordance with ASTM E 1355, *Standard Guide for Evaluating the Predictive Capability of Deterministic Fire Models* [Ref. 1]. MAGIC is a zone model developed and maintained by Electricité de France (EdF), which officially released the latest version of the model (Version V4.1.1b) in 2005. The MAGIC software calculates fire-generated conditions (e.g., hot gas layer temperature, etc) in single- or multi-compartment geometries as a function of time [Refs. 2, 3, and 4].

The MAGIC software is a classical two-zone model for fire simulations, with capabilities to process multi-compartment problems. Each compartment is divided into two volumes, which are assumed to have homogeneous thermo-physical properties. The solution of the mass and energy balances accumulated in each zone, together with the ideal gas law and equation of heat conduction into the walls, results in the predicted environmental conditions generated by the fire.

Consistent with ASTM E 1355, this document is structured as follows:

- Chapter 2 provides qualitative background information about MAGIC and the V&V process.

- Chapter 3 presents a technical description of MAGIC, which includes the underlying physics and chemistry inherent in the model. The description includes assumptions and approximations, an assessment of whether the open literature provides sufficient scientific evidence to justify the approaches and assumptions used, and an assessment of empirical or reference data used for constant or default values in the context of the model. MAGIC's source code and technical description are EdF proprietary material (available to EPRI members only); consequently, this report provides only a technical summary of this material.

- Chapter 4 documents the mathematical and numerical robustness of MAGIC, which involves verifying that the implementation of the model matches the stated documentation.

- Chapter 5 presents a sensitivity analysis. The sensitivity analysis explores and discusses the effects of variations in the input parameters on MAGIC outputs.

- Chapter 6 presents the results of the validation study in the form of relative differences classified by fire modeling attribute. The following attributes were selected for validation purposes:

  - hot gas layer temperature and height
  - ceiling jet temperature
  - plume temperature
  - flame height
  - oxygen concentration
  - smoke concentration
  - room pressure
  - target surface temperature and incident radiant and total heat flux
  - wall surface temperature and incident total heat flux

- Appendix A presents the technical details supporting the calculated relative differences discussed in Chapter 6 and provides graphical comparisons of experimental measurements and modeling results.

- Appendix B presents MAGIC input files.

2
MODEL DEFINITION

This chapter provides qualitative background information about MAGIC and the V&V process, as suggested by ASTM E 1355.

2.1 Name and Version of the Model

This V&V study is for MAGIC Version V4.1.1b, which EdF released in November 2005.

2.2 Type of Model

MAGIC is a two-zone fire model that predicts the environmental conditions resulting from a fire prescribed by the user within a compartmented structure. Essentially, the space to be modeled is subdivided into two control volumes that represent upper and lower layers. The fundamental equations for conservation of energy and mass are solved in each control volume as the fire heat release rate develops over time.

2.3 Model Developers

MAGIC was developed and is maintained by Electricité de France (EdF).

2.4 Relevant Publications

MAGIC is supported by three EdF publications, including (1) the technical manual, which provides a mathematical description of the model [Ref. 2]; (2) the user's manual, which details how to use the graphical interface [Ref. 3]; and (3) the validation studies, which compare MAGIC's results with experimental measurements [Ref. 4]. These three proprietary publications are available through EPRI to EPRI members.

2.5 Governing Equations and Assumptions

MAGIC solves the conservation equations for mass and energy. The model does not explicitly solve the momentum equation, except for use of the Bernoulli equation for the flow velocity at room openings. These three equations and the ideal gas law are solved to obtain fire-generated conditions in the selected control volumes.

MAGIC assumes that the room is divided in two zones (upper and lower control volumes), in which the equations described above are solved. The upper control volume, referred to in this report as the hot gas layer, is assumed to have uniform density and, therefore, temperature. The same assumption applies to the lower control volume (also known as the lower layer).

Chapter 3 of this report and Reference 2 provide a complete technical description of MAGIC algorithms and sub-models.

2.6 Input Data Required To Run the Model

In general, the following data are necessary to develop the input file for MAGIC. The required inputs for each individual analysis may vary, and depend on the characteristics and objectives of the fire scenario under analysis.

(1) Parameters describing the compartment geometry and ventilation conditions:

- Compartment geometry (length, width, and height): The compartment (or each compartment in a multi-room scenario) is assumed to have a rectangular floor base and flat ceiling.

- Floor, ceiling, and wall material properties (density, specific heat, and thermal conductivity): Depending on the selected material, this information may be available in the MAGIC database.

- Natural ventilation (height and width of doors; height, width, and elevation of windows; time to open/closed doors and windows during a fire simulation; and leakage paths).

- Mechanical ventilation (injection and extraction rates, vent elevations, and time to start/stop the system).

(2) Parameters describing the fire characteristics:

- Fuel type and fire heat release rate profile, which is specified using the heat of combustion and the mass loss rate of the fuel

- Fire location (elevation, near a wall, near a corner, or center of room)

- Footprint area of the fire: circular (e.g., pool fires specified by the diameter) or rectangular (e.g., bounded pool fires, electrical cabinets specified by length and width)

- Fuel mass, irradiated fraction, and stochiometric fuel-oxygen ratio

(3) Two sets of parameters (thermo-physical properties and location) describe targets. Thermo-physical properties include the density, specific heat, and thermal conductivity of the material. Location refers to where the target is with respect to the fire (expressed with three-dimensional coordinates).

(4) The inputs for sprinklers and detectors are the device's location with respect to the fire and its response characteristics, which include activation temperature and response time index.

The MAGIC user's guide [Ref. 3] provides a complete description of the input parameters required to run MAGIC.

2.7 Property Data

Various equations associated with the MAGIC model require the following property data:

- For walls: density, thermal conductivity, and specific heat
- For targets: density, thermal conductivity, and specific heat
- For fuels: heat of combustion, mass loss rate, stochiometric fuel-oxygen ratio, specific area, and radiated fraction

These properties may be available in fire protection engineering handbooks or the MAGIC database. However, depending on the application, properties for specific materials may not be readily available.

2.8 Model Results

MAGIC has an extensive library of output values. Once a given simulation is completed, MAGIC generates an output file with all of the solution variables. Through a "post-processor" interface, the user selects the relevant output variables for the analysis. Typical outputs include (but are not limited to) the following examples:

- environmental conditions in the room (such as hot gas layer temperature, oxygen concentration, and smoke concentration)

- heat transfer-related outputs to wall and targets (such as incident, convective, radiated, and total heat fluxes)

- oxygen effects on heat release rate and flame height

- flow velocities through vents and openings

3
THEORETICAL BASIS FOR MAGIC

3.1 Introduction

This chapter provides a brief technical summary of the MAGIC zone model to address
the ASTM E 1355 guidance to "verify the appropriateness of the theoretical basis
and assumptions used in the model." However, given the proprietary nature of the software,
readers should refer to Reference 2 for a complete technical description.

3.2 Theoretical Basis for MAGIC

MAGIC is a classical two-zone fire model. That is, a room is divided into upper and lower zones
(or layers). The upper layer (also referred to as the hot gas or smoke layer) accumulates hot gases
generated in the combustion zone and primarily transported by the fire plume. The lower layer
primarily consists of fresh air and has its own energy and mass balance.

Perhaps the most important characteristic of the two-zone model formulation is that each zone
is assumed to have homogeneous thermo-physical properties. The gas density (and, consequently,
the temperature), oxygen concentration, and concentration of unburned gases are assumed to
remain constant throughout each layer. These properties change only as a function of time.

The resulting fire conditions are obtained by solving equations for conservation of mass, species,
and energy, together with the ideal gas law. The species equation yields the concentration
of unburned fuel and oxygen in each layer. The compartment pressure, layer temperature,
and layer heights are obtained from the mass and energy equation. Finally, the gas densities
are calculated using the ideal gas law.

MAGIC provides the following general results:

- temperatures of hot and cold zones
- concentrations of oxygen and unburned gases
- smoke migration into each room
- the mass flow rates of air and smoke through the openings and vents
- the pressures at the floor level of each room
- the temperatures at the surface of the walls
- the thermal fluxes (radiative and total) exchanged by the targets placed by the user

From a geometric point of view, MAGIC works on a set of rectangular rooms with flat ceilings,
with their edges parallel to the reference axes. These rooms communicate with each other and
the outside through horizontal or vertical openings. Figure 3-1 summarizes MAGIC's modeling
features.

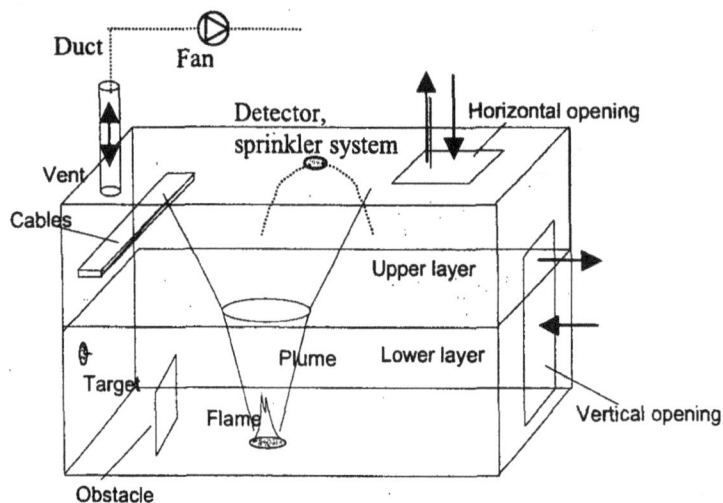

Figure 3-1: Pictorial Representation of MAGIC's Features

3.2.1 Combustion

The standard combustion model in MAGIC assumes a perfect oxidation reaction; that is, the fire will burn at the specified heat release rate if oxygen is available. MAGIC tracks the amount of oxygen in the fuel (in the case of a pre-mixed fuel), oxygen entrained by the fire, unburned fuel in the environment, and the predefined fuel source in order to determine whether complete combustion will occur. The chemical aspects of combustion are not considered. If the oxygen entrained into the plume is at least equal to the quantity necessary to burn all of the gaseous fuels in the plume, combustion is considered complete and controlled by the fuel flow rate. If not, the combustion is incomplete and controlled by the available oxygen. The user can also specify a low oxygen limit (LOL).

3.2.2 Hot Gas Layer Temperature and Height

Hot gas layer temperature and height result from the conservation equations of energy and mass for the defined control volume. Properties are assumed to be homogeneous in the volume except in the specific regions of the plume and ceiling jet. Mass balance takes into account the fire plume flow from the lower layer, and air supplied or exhaust through vents or openings. Energy balance takes into consideration convection and radiation to the room surfaces (walls ceiling and floor) and to the lower layer. The radiation properties of the layer are obtained from its opacity (based on smoke concentration resulting from the mass balance). Oxygen and unburned gas concentrations also result from the mass and energy balances in the hot gas layer volume. Similar conservation equations are applied to the lower layer.

3.2.3 Walls, Ceiling, and Floor

Walls, ceiling, and floor are represented using one-dimensional finite difference meshing of conduction. Two separate calculations are made, with one for the section of wall in the upper layer and the ceiling and a similar one for the lower layer and the floor. Boundary conditions for walls inside a room use convection and a detailed radiation exchange. As a default, heat transfer coefficient and wall emissivity are fixed to 15 W/m^2-K and 0.9, respectively. The heat transfer coefficient can also be correlated to the temperature and the estimated velocity in the layer, as an option. This study is based on default values.

Each wall can be constructed with multiple successive layers of a homogenous material; however, the characteristics of each material are assumed to remain constant. The initial temperature condition at both sides of the wall is the ambient temperature. The boundary conditions are calculated as the simulation goes on and are based on heat exchange between surfaces and gas layers.

3.2.4 Flame Height, Fire Plume & Ceiling Jets

The fire plume in MAGIC is modeled using McCaffrey's semi-empirical correlations for fire plume entrainment [Ref. 5], McCaffrey's correlation for temperature and velocity in the flame region [Ref. 5], Heskestad's correlation for temperature and velocity in the plume region [Ref. 6]. The software incorporates the effects of the smoke layer on fire plume temperature [Ref. 6], and simulates ceiling jets using the model developed by Cooper [Ref. 7] to account for hot gas layer effects. As such, MAGIC models both confined and unconfined ceiling jets and considers the adiabatic ceiling jet correlation and exchanges to walls from the layers' properties. In addition, MAGIC accounts for fires located along a wall or in corners, and it estimates flame height using Heskestad's correlation [Ref. 7].

3.2.5 Natural & Mechanical Ventilation

Horizontal Openings: The model for flows through horizontal (ceiling or floor) openings is based on the formulation proposed by Cooper [Ref. 8]. This model addresses the issue of one- or two-way flow at the opening using experimental results. It is important to note that this model has been developed from experimental conditions in which the horizontal opening was not directly above the fire source. *The model does not apply to configurations in which the fire plume directly influences the flow.*

Vertical Openings: MAGIC uses the Bernoulli equation to model flows through vertical openings with a corresponding orifice flow coefficient. Flows are assumed to be perpendicular to the surface of the opening.

The ventilation model used in MAGIC is represented by a fan between ducts. In each duct, regular and singular pressure differences are considered. Upstream and downstream nozzles make the link between rooms and vent systems. In the case of no fan, the model calculates the mass flow through the ducts considering pressure differences.

3.2.6 Heat Transfer

Radiation: Radiation modeling is relatively complex in MAGIC. The gas layer is treated as a semi-transparent gas. Radiation exchanges between surfaces (walls and openings), flames, and gas layers are considered. One system is built for the upper layer and another for the lower layer. View factors are re-evaluated for each iteration as a result of the layer interface height variations. Specific configuration factors are also calculated for targets and cables.

Convection: For the case of convection to room surfaces, the heat transfer coefficient and wall emissivity are fixed to 15 W/m^2-K and 0.9, respectively. The heat transfer coefficient can also be correlated to the temperature and the estimated velocity in the layer, as an option. This study is based on default values.

In the case of targets, the convective exchanges use a variable coefficient of exchange, h. This coefficient has two components, one of which is attributable to natural convection and the other to forced convection. These two terms are calculated from correlations listed in Reference 2.

Conduction: Walls, ceiling, and floor are represented using one-dimensional finite difference meshing of conduction. Two separate calculations are made, with one for the section of wall in the upper layer and the ceiling and a similar one for the lower layer and the floor. Each wall can be constructed with multiple successive layers of a homogenous material; however, the characteristics of each material are assumed to remain constant. The initial temperature condition at both sides of the wall is the ambient temperature. The boundary conditions are calculated as the simulation goes on and are based on heat exchange between surfaces and gas layers.

This modeling approach also applies to heat conduction into cable targets. Slab thermal targets however are specified with a single layer of material.

3.2.7 Smoke Concentration

The relevant output for smoke concentration in MAGIC is the average extinction coefficient, k, with units of 1/m. The relevant input value governing k is the specific area, s (units of m^2/g), which is calculated using the soot yield of the fuel, y_s, as follows $s = y_s k_m$ where k_m is a constant value of 7,600 m^2/g [Ref. 21]. The average extinction coefficient can be converted to concentration in units of mg/m^3 or visibility in units of m with relatively simple algebraic manipulations. For the purpose of NPP applications, visibility would be the most relevant output. Recall from reference 21 that the average extinction coefficient correlates linearly with visibility, based on the equation $S = 3/k$ for a light-reflecting object, or $S = 8/k$ for a light-emitting object, where S is the visibility distance in m.

The average extinction coefficient is converted into smoke concentration using the equation:

$$\upsilon = \frac{k}{k_m}$$

where υ is the concentration in mg/m^3, and k_m is a constant with value 0.0076 m^2/mg [Ref. 21].

3.2.8 Targets

Two kinds of targets are implemented in MAGIC. The basic target is equivalent to a flux meter (with controlled surface temperature), and the thermal target is equivalent to a one-dimensional homogeneous material. Fire sources, gas layers, walls, and openings generate the incident heat fluxes to the targets. Both incident convective and radiative fluxes are considered. For example, MAGIC considers direct radiation flux from sources located in adjacent rooms, and correlates the convective exchange to local temperature and gas velocity. The target can be located in the plume or ceiling jet, and MAGIC calculates the target temperature using a one-dimensional finite difference conduction model throughout the thickness of the target.

3.2.9 Electrical Cables

Electrical cables serve as both fuel and targets in NPP fire scenarios. Figure 3-2 summarizes the modeling of electrical cables in MAGIC.

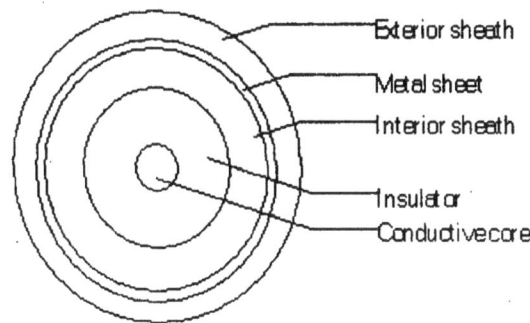

Figure 3-2: Simplified Inner Structure of an Armored Electrical Cable

The cable is composed of successive layers of user-specified materials. Each layer includes a certain number of discretization points, limited to 40 per layer. The code automatically implements a node at the center of the cable, and the heat transfer inside the cable is considered to follow an axial symmetry.

In the calculation, an electrical cable is divided in 20-cm (7.9-inch) segments along its length. The total number of segments should not exceed 50. For each segment, MAGIC calculates the thermal exchanges with the outside and the thermal heating.

The maximum surface temperature encountered on all the segments is the criterion to start the cable ignition, from a piloted ignition threshold value or (if needed) a pyrolysis output (introduced by the user). After the ignition, the cable behaves as a classical fuel, and its thermal behavior is no longer modeled (that is, the surface temperature retains its last value).

An important consideration in this validation study is the treatment of multi-conductor cables. In MAGIC, multi-conductor cables were modeled as single-conductor cables, as follows:

- The cross-sectional area of the equivalent single-conductor is $\sum_{i=1}^{n} A_i$, where Ai is the area of each individual conductor, and n is the total number of conductors in the cable.

- The thickness of the jacket remains the same.

- The thickness of the insulation is given by (cable thickness – jacket thickness – equivalent conductor radius).

Figure 3.3 illustrates this process.

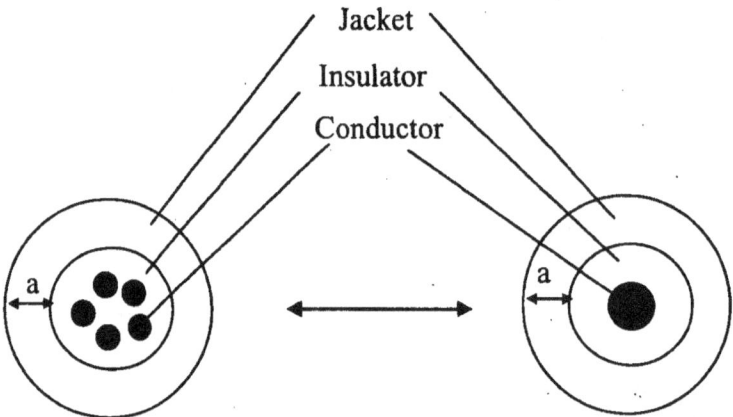

Figure 3-3: Modeling Multi-Conductor Cables in MAGIC

3.2.10 Sprinkler Suppression

Modeling of sprinkler suppression is divided into three phases:

(1) Sprinkler activation determines the instant when the device is activated. Specifically, the sprinkler is triggered when the temperature of the gas contained in the sprinkler bulb reaches its activation temperature, which generally varies from 70 to 150°C (158 to 302 °F) depending on the sprinkler. The model for sprinkler activation developed by Heskestad [Ref. 9] is implemented in MAGIC.

(2) Cooling of the hot gas layer by the water spray is achieved through the interaction between water droplets and the hot gas layer, which results in several physical and thermal phenomena. The spray comprises a multitude of drops that have different speeds, diameters, and directions. The thermal exchanges between the hot gases and the drops increase the temperature of the drops and lead to partial or total evaporation and cooling of hot gases.

(3) Fire extinction takes into account the interaction between water drops and the fire. As a conservative approach, the heat release rate will remain at the intensity it had at the time of sprinkler activation.

3.3 Concluding Remarks

This chapter provided an overview of the modeling features of MAGIC. A complete technical description is available in Reference 2.

MAGIC is based on a combination of macroscopic conservation equation and empirical correlations for specific phenomena. This combination between fundamental principles and experimental observations leads to a sound quantitative approach for its intended domain of application. In addition to the validity of semi-empirical sub-models that may be used independently, confidence in the predictive capabilities of the code is mainly obtained through validation exercises, were the sub-models work together in order to provide consistent results for the different relevant outputs in a specific scenario.

Although MAGIC can be used for general fire modeling applications, it has always been intended for nuclear power plant applications. For this reason, its validation file and some of its sub-models (e.g., electrical cables) are specially adapted to this field.

It is necessary to stress the importance of the input parameters. The databases in MAGIC give consistent information to the user, who can also customize with scenario specific materials. MAGIC also provides some checks for the validity of the input values. However, typical fire modeling studies usually involve uncertain inputs. In those cases, the analyst's expertise and experience in the field is important for developing valid input files and obtaining consistent conclusions from the model results.

MAGIC includes an extensive list of output values. The outputs are classified in the following groups: Room, Wall, Targets, Source, Opening, Vents, Furniture, and Others. A number of output options can be found within each group.

Only a selected number of output options were subjected to the V&V study. Table 3-1 lists such output options. The outputs are labeled as "Group/output option". For example, the hot gas layer temperature can be found in the MAGIC post-processor under "Room/Upper layer temperature".

Table 3-1: MAGIC Capabilities Included in the V&V Study.

Model Capability	MAGIC output V&V	Comment
Hot gas layer temperature	Room/Upper layer temperature	Other layer temperature options, e.g. lower layer temperature, were not subjected to V&V.
Hot gas layer height	Room/layer interface height	
Plume temperature	Target/Gas temperature (for a target sensor located in the fire plume)	
Ceiling jet temperature	Target/Gas temperature (for a target sensor located in the ceiling jet)	
Target temperature	Target/Gas temperature Target/Surface temperature of the target	
Wall temperature	Target/Surface temperature of the target	MAGIC outputs under WALL classifications were not subjected to V&V

Model Capability	MAGIC output V&V	Comment
Target heat flux	Target/Incident heat flux (for radiation) Target/Total heat flux (for total heat flux)	
Wall heat flux	Target/Total heat flux	MAGIC outputs under WALL classifications were not subjected to V&V
Room pressure	Room/Pressure at the room floor	
Flame height	Source/Height	Other output options under the SOURCE classification were not subjected to V&V.
Smoke concentration	Room/Upper layer extinction coefficient	Other layer extinction coefficient options, e.g. lower layer extinction coefficient, were not subjected to V&V.
Oxygen concentration	Room/Upper layer oxygen concentration	Other layer concentration options, e.g. lower layer oxygen concentration, were not subjected to V&V.
Fire suppression	Fire suppression features and effects in MAGIC were not subjected to V&V.	
Furniture (Obstructions)	Furniture (Obstructions) features and effects in MAGIC were not subjected to V&V.	
Fire extinction	Fire extinction capabilities in MAGIC were not subjected to V&V.	
Cable temperature	The Cable feature in MAGIC was not subjected to V&V.	
Cable heat flux	The Cable feature in MAGIC was not subjected to V&V.	

Model Capability	MAGIC output V&V	Comment
Vent flows	Vent flows modeling capabilities in MAGIC were not subjected to V&V.	
Flow velocities	Vent flows modeling capabilities in MAGIC were not subjected to V&V.	

4
MATHEMATICAL AND NUMERICAL ROBUSTNESS

4.1 Introduction

This chapter documents the mathematical and numerical robustness of MAGIC, which involves verifying that the implementation of the model matches the stated documentation. Specifically, ASTM E 1355 suggests the following analyses to address the mathematical and numerical robustness of models:

- Analytical tests involve testing the correct functioning of the model. In other words, these tests use the code to solve a problem with a known mathematical solution. However, there are relatively few situations for which analytical solutions are known.

- Code checking refers to verifying the computer code on a structural basis. This verification can be achieved manually or by using a code-checking program to detect irregularities and inconsistencies within the computer code.

- Numerical tests investigate the magnitude of the residuals from the solution of a numerically solved system of equations (as an indicator of numerical accuracy) and the reduction in residuals (as an indicator of numerical convergence).

4.2 Mathematical and Numerical Robustness Analyses for MAGIC

MAGIC consists of a user interface and a mathematical source code models. This section covers only the second module in detail. Section 4.2.4 describes a classical quality assurance policy for the user interface.

4.2.1 Comparison with Analytical Solutions

General analytical solutions do not exist for most fire problems. Nonetheless, it is possible to test specific aspects of the model in typical situations. Some studies have been performed to control the correct behavior of the following sub-models of MAGIC:

- conduction into the wall: comparison to other models and analytic solutions

- target and cable thermal behavior: consistency of the behavior in typical situations

- plumes model: comparison with the theoretical model

- vent and opening: comparison to other zone and field models

- room pressure: comparison with pressure estimated by the perfect gas law and simplified energy equation

These studies are EdF's proprietary material.

4.2.2 Code Checking and Code Quality

In general, MAGIC is structured as shown in Figure 4-1. The code reads data in a case file (*.cas, which can be accessed using any word processing software or the MAGIC interface itself), and initializes all of the variables for the problem. To solve the system of differential equations, the model divides time into successive intervals, and then solves the equations recursively from instant t_0 (where the variables are known) and by using a recurrence formula linking instant t_n to t_{n+1}.

All the solution variables at instant t_n are obtained through Array Y (temperatures, pressure, concentrations, and gas characteristics). The 26 constituent equations related to a given room are numbered from Y(1) to Y(26).

A subroutine calculates all of the fluxes (Y') derived from the physical model implementation between t_n and t_{n+1}. This Array, Y', is transmitted to the solver for the calculation of t_{n+1}. The ordinary differential equation (ODE) solver is based on a trial-and-error process to estimate the solution variables at t_{n+1}.

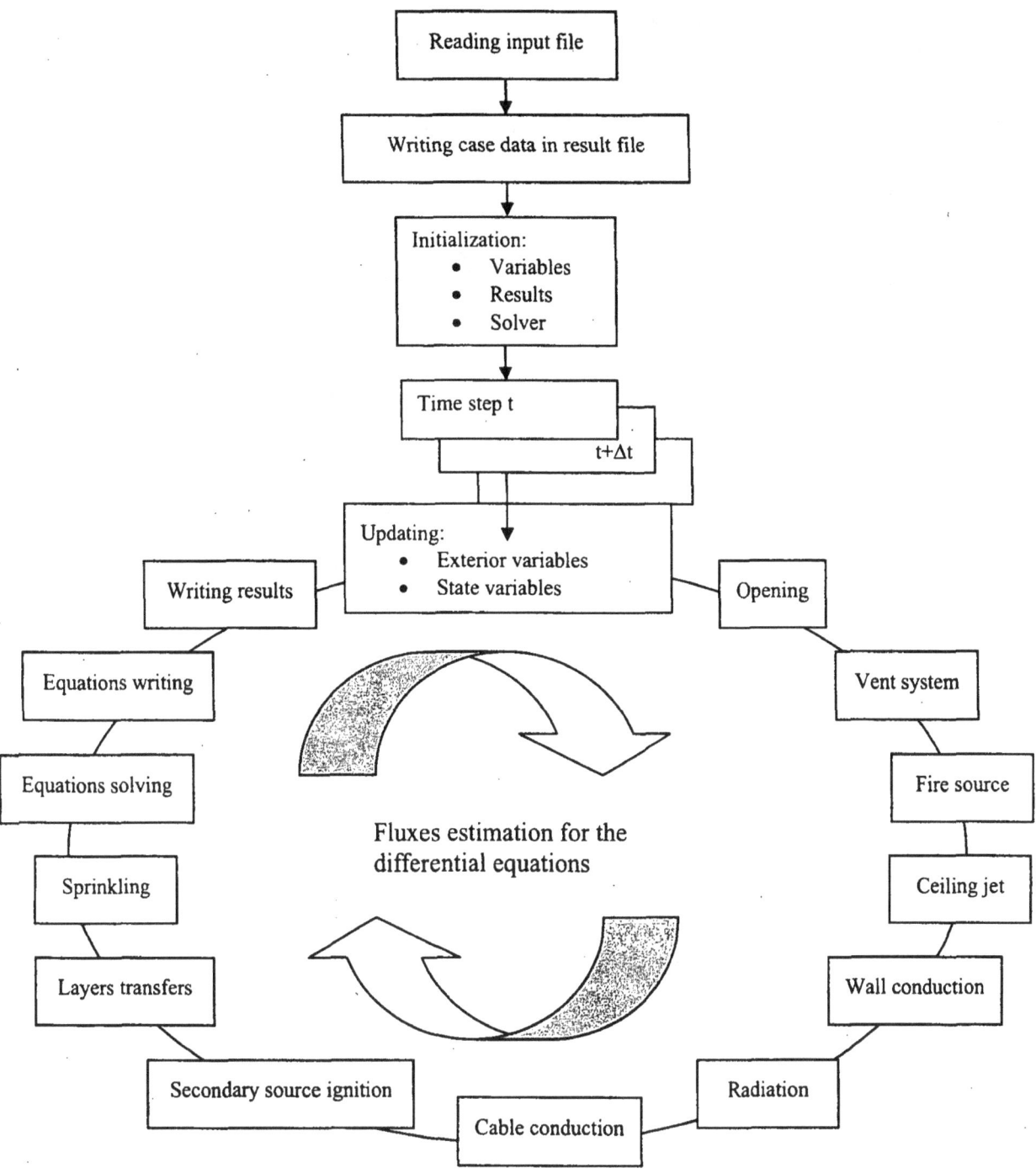

Figure 4-1: Simplified Functional Breakdown of the MAGIC Code

The equations of conduction inside the walls and cables, the sprinkler spray, and the equations of transport in the ventilation system are solved independently during the calculation cycle of transfers. All temperatures are updated at each calculation step.

The ODE system is solved using backward differential formulas (Gear method). The solver uses a specific algorithm based on the BROYDEN approach [Refs. 10 and 11]. This method is interesting because there are several distinct time scales in the resolution of the ODE. Indeed, the fire combustion is the fastest phenomenon encountered, and the transport time scale is much slower than the fire reaction. This is why a numerical resolution (such as Gear) with trial-and-error enables the model to dynamically adjust the resolution time step. If the problem presents dramatic physical changes in relatively short periods of time, the time step decreases; however, in the opposite case, the time step increases.

In addition, the BROYDEN approach is a quasi-Newton method to process the system of nonlinear algebraic equations. In this method, the Jacobean matrix is replaced by a series of "approached" matrices converging toward the exact matrix at the solution point. First, the "approached" matrix is decomposed in a product of two matrices — LU with L as the lower triangular matrix, and U as the upper triangular matrix with 1 on the diagonal (CROUT method).

The resolution makes a first iteration with the Newton method, and three iterations with the BROYDEN method. This solution enables the model to improve the convergence when the problem presents dramatic physical changes in relatively short periods of time. This method avoids recalculation of the Jacobean matrix in each iteration (thereby saving time).

The source code itself is tested with the following methods:

- First, to control robustness, the code may be compiled in several different platforms and software applications. The MAGIC code has been compiled under Microsoft® Windows 2000® and Windows XP®, with a variety of compilers, including Absoft Pro FORTRAN, Visual FORTRAN, and G77. In addition, a global update of the FORTRAN sources was performed in 2004 [Ref. 12], and aspects such as code documentation, variable glossary, and source cleanup were addressed.

- In terms of code quality, two tools have been used to control the language:

  ➢ FOR_STUDY from Cobalt Blue [Ref. 13]

  ➢ plusFORT from Polyhedron Software [Ref. 14]

These tools confirm the consistency of variables and constants (undefined and incorrectly or redundantly declared) and use of good FORTRAN syntax.

The software quality assurance system [Ref. 15] provides a process to fix detected anomalies concerning the interface of the code. Maintenance of MAGIC is based on observation forms, which identify problems. Then, a modification form describes the problem analysis and proposed solutions. Finally, a correction form explains the chosen solution and implementation features. The project manager decides on the implementation of the correction in future versions.

4.2.3 Numerical Tests

For each new code version, a set of tests is used to ensure that the calculation is correct. These tests come from previous case studies. The convergence and speed of the calculation is the first step of control. Main results from the original study are then compared, and significant differences are analyzed. These studies are EdF proprietary material.

Specific tests are performed in the maintenance process when new models are implemented into the code, or when existing models are corrected or improved. Those tests are not systematically conducted for new versions, but they are available in case problems arise with the model under study. The tests are mentioned in the correction report [Ref. 15] which is kept for each code correction. These reports are EdF proprietary material and not published.

4.2.4 User Interface

The method used is a classical V-cycle development with tests, as illustrated in Figure 4-2.

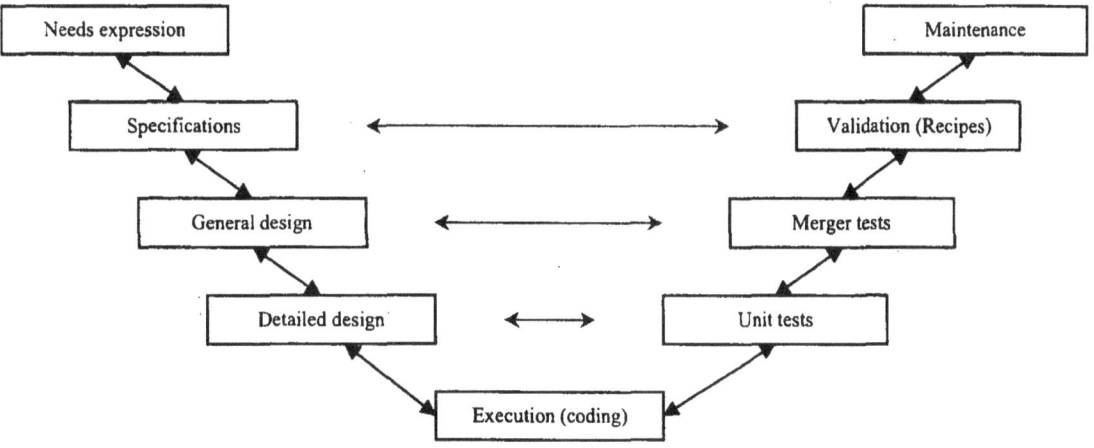

Figure 4-2: V-Cycle Representation

The code meets the corresponding specifications at each step of the cycle. The following reference documents are available:

- Conception documentation [Ref. 16] presents the general "architecture" and input/output files, and summarizes all of the class, function, and subroutine codes in the interface. The document includes a brief description of objectives and parameters for each.

- Reception test framework [Ref. 17] validates all of the interface functions and enables the user to verify consistency between the specification and software.

- User reference guide [Ref. 18] presents the various interface menus and details their uses.

4.3 MAGIC Improvements as a Result of the V&V Process

Some improvements were made in MAGIC as a result of the V&V process. This highlights the importance of the V&V process, including a rigorous comparison of code predictions with experimental observations. MAGIC Version 4.1.1b includes the latest the improvements resulting from the V&V. Specifically, the following MAGIC features were corrected during the V&V process:

- improvement of the soot mass balance within the plume
- improvement of the correlation for temperature in the flame region
- correction of a problem in the flame length calculation when lower than the layer interface

The updated features are completely described in Reference 2.

4.4 Concluding Remarks

MAGIC has been developed to allow quick and robust calculations of typical fire conditions in single- and multi-compartment building, on a standard personal computer platform. Calculations will be very quick (a few seconds) for simple scenarios (e.g., single room with opening and vents). Configurations with several communicating rooms (up to 24) can be managed by the code. Calculation times however are correlated to the complexity of the problem. The number of communicating rooms maybe the most influencing parameter. The use of the cable model can also have a significant "cost" on calculation time.

The development and maintenance of MAGIC is performed by EdF Research and Development (R&D). On average, a MAGIC revision is released once a year.

5
MODEL SENSITIVITY

This chapter discusses the MAGIC sensitivity analysis, which ASTM E 1355 defines as a study of how changes in model parameters affect the results. In other words, sensitivity refers to the rate of change of the model output with respect to input variations. The purpose of this sensitivity analysis is twofold:

1. Test MAGIC's predictive capabilities with a range of different inputs to check for consistency in the results.

2. Compare different modeling strategies in MAGIC in support of the validation study described in Chapter 6 and Appendix A. Specifically, the following two modeling strategies were selected for validation:

   a. use of thermic-target (slab) sub-model in MAGIC for predicting cable surface temperature versus the use of the cable sub-model

   b. use of thermic-target (slab) sub-model in MAGIC for predicting temperature and heat fluxes to room surfaces versus the use of the wall temperature sub-model

5.1 Definition of Base Case Scenario for Sensitivity Analysis

Conducting a sensitivity analysis requires the definition of a base case scenario. Variations in the output of the model are measured with respect to the base case scenario.

The base case scenario for this study was analyzed in Benchmark Exercise (BE) #1, as part of an ongoing International Collaborative Fire Modeling Project (ICFMP) [Ref. 19]. This section summarizes the technical description of the scenario. (Note that only Part 1 of BE #1 was selected as the base case.)

- Room
  - Length: 15.2 m (50 ft)
  - Width: 9.1 m (30 ft)
  - Height: 4.6 m (15 ft)
  - Walls: 0.15 m thick concrete (6 in)
  - Door: 2.4 m x 2.4 m (62 ft^2)
  - Mechanical ventilation: 5 air changes per hour
  - Vent size: 0.5 m^2 (5.4 ft^2)
  - Vent elevation: 2.4 m (7.9 ft)

- Target
  - Cable Tray A: 0.6 m wide, 0.08 m deep
  - Elevation (cable tray A): 2.3 m, 0.9 m off the right wall of the room
  - Cable Tray B: 0.6 m wide, 0.08 m deep
  - Elevation (Cable Tray B): 2.3 m, along the left wall of the room

- Material properties for concrete
  - Specific heat: 1,000 J/kg-K
  - Thermal conductivity: 1.75 W/m-K
  - Density: 2200 kg/m3
  - Emissivity: 0.95

- Material properties for cables (targets): See Table 5-1.

Table 5-1: Material Properties for Cables

Material	Thermal cond [kW/mK]	Density [kg/m3]	Cp [kJ/Kg-k]
XPE	0.00021	1375	1.566
PVC	0.000147	1380	1.469

- Ambient conditions
  - Temperature: 27 °C (81 °F)
  - Relative humidity: 50%
  - Pressure: 101,300 Pa
  - Elevation: 0
  - Wind speed: 0

- Fire (heat release rate): The fire heat release rate was assumed to have a t^2 growth profile. The fire reaches its peak intensity in 600 seconds. Two peak intensities (1.0 MW and 5.0 MW) were selected for this sensitivity analysis, in order to explore different MAGIC features and capabilities. For example, fire intensity capable of consuming all of the oxygen in the enclosure allows the sensitivity analysis to explore situations that exercise MAGIC's extinction model. Figure 5-1 illustrates the two heat release rate profiles.

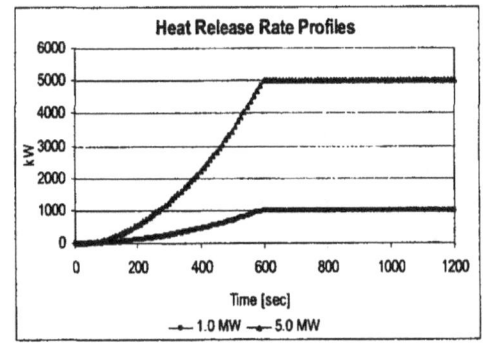

Figure 5-1: Selected Heat Release Rate Profiles

5.2 Sensitivity Analysis

A total of 16 MAGIC simulations were conducted for this sensitivity analysis. A key input parameter was modified in each run in order to explore MAGIC's capabilities and sensitivities with respect to input parameters. Note, however, that MAGIC requires numerous input parameters, and this study did not analyze all of those input parameters and their combinations. Table 5-2 summarizes the 16 MAGIC fire simulations selected for sensitivity analysis.

Table 5-2: Summary of MAGIC Simulations Selected for Sensitivity Analysis

Case	Heat Release Rate [kW]	Natural Ventilation [m²]	Mech. Ventilation [m³/s]	Lower Oxygen Limit [%]	Fuel Type	Vertical Fire Position	Horizontal Fire Position
1	1000	0.015	0	0	Heptane	Floor	Center
2	1000	0.015	0.88	0	Heptane	Floor	Center
3	1000	0.015	0	10	Heptane	Floor	Center
4	1000	0.015	0.88	10	Heptane	Floor	Center
5	1000	5.76	0	0	Heptane	Floor	Center
6	1000	5.76	0	0	Toluene	Floor	Center
7	1000	5.76	0	0	Heptane	0.5H	Center
8	1000	5.76	0	0	Heptane	Floor	0.25W
9	5000	0.015	0	0	Heptane	Floor	Center
10	5000	0.015	0.88	0	Heptane	Floor	Center
11	5000	0.015	0	10	Heptane	Floor	Center
12	5000	0.015	0.88	10	Heptane	Floor	Center
13	5000	5.76	0	0	Heptane	Floor	Center
14	5000	5.76	0	0	Toluene	Floor	Center
15	5000	5.76	0	0	Heptane	0.5H	Center
16	5000	5.76	0	0	Heptane	Floor	0.25W

The first eight simulations were conducted with the assumption of a peak fire intensity of 1.0 MW. The researchers varied parameters affecting the size of the ventilation openings, the mechanical ventilation system, the fuel type, and the fire location, in order to explore their effects on selected results. Simulations 9–16 are identical to the first eight, but with a peak fire intensity of 5.0 MW. Targets of both PVC and XPE material were specified in the computational domain. Therefore, sensitivities to thermo-physical properties of targets can be explored in each of the analyzed cases. Figure 5-2 provides a pictorial representation of the fire scenario selected for sensitivity analysis, as defined in MAGIC.

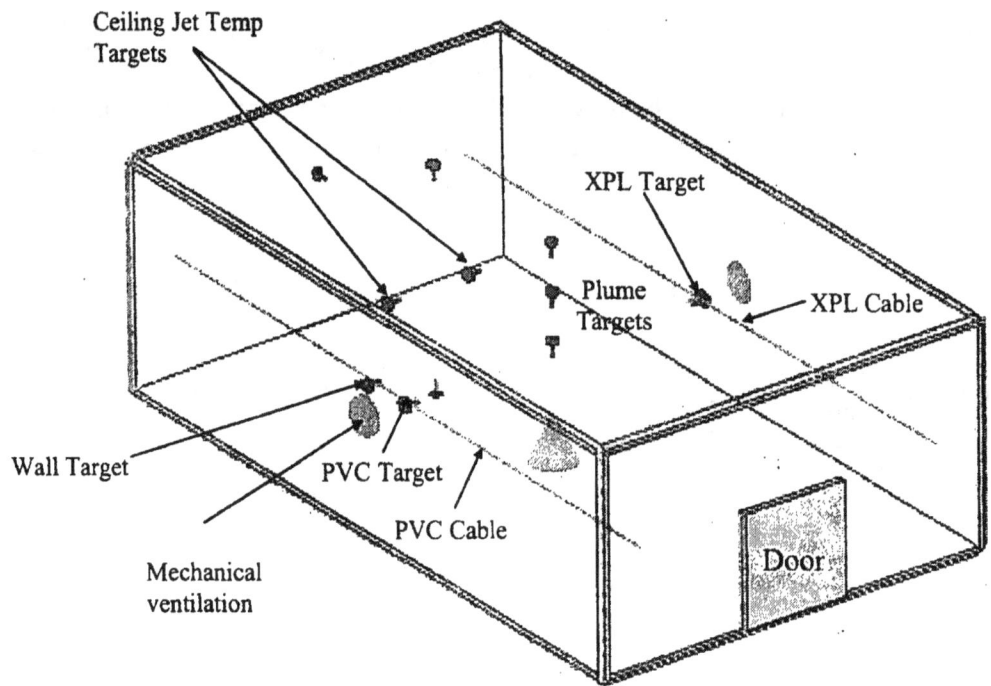

Figure 5-2: Problem Specification in MAGIC

The sensitivity analysis and its results are classified by relevant fire modeling attributes selected for this V&V study, as presented in the following sections.

5.2.1 Hot Gas Layer Temperature and Height

The hot gas layer temperature is perhaps the single most important output of a zone model, because it is the direct result of the energy and mass balance in the upper control volume. In general, the hot gas layer temperature is affected by the fire intensity, natural and mechanical ventilation characteristics, and material properties of the room. This study analyzed the effect of the first three groups of inputs on the hot gas layer temperature. The properties of the wall were not evaluated, because most NPP rooms have concrete floors, ceilings, and walls.

Figure 5-3 summarizes the hot gas layer temperature profiles for selected cases in the sensitivity analysis. The first group of profiles (Cases 1, 2, and 5) was associated with a heat release rate of 1.0 MW, and the predicted hot gas layer temperature was just below 140 °C (284 °F). Notice that the temperature profile is similar to the heat release rate. That is, once the fire reaches steady-state at 600 seconds, the temperature profile is almost steady.

The second group of profiles (Cases 9, 10, and 13) reached temperatures just below 350 °C (662 °F). Notice that the profiles for Cases 9 and 10 show decay after 600 seconds, which is attributed to a reduction in the heat release rate as a result of low oxygen concentration. Recall that only air leakages were assumed in these two simulations. Notice that Case 13 was not affected by the amount of oxygen because the door was open and fresh air was constantly moving into the enclosure.

In summary, although it is generally obvious that the heat release rate affects the hot gas layer temperature, higher fire intensities consume additional oxygen, which may prevent the fire from burning at its specified heat release rate.

The hot gas layer height output is directly associated with the hot gas layer temperature, as it is also a direct output of the energy and mass balance in the upper control volume. This output result is also generally affected by the same input parameters as the hot gas layer temperature. Figure 5-3 also illustrates a selected set of hot gas layer heights. Notice two distinctive sets of results. First, the profiles for Cases 1, 2, 9, and 10 reached the floor of the room. Those cases consist of fire simulations that assume a small leakage area below the door (closed door simulation). In Cases 5 and 13, the hot gas layer did not reach the floor, because the door was assumed to be open. As expected, the layer interface in Case 13 leveled lower than the one in Case 5, because of the higher heat release rate.

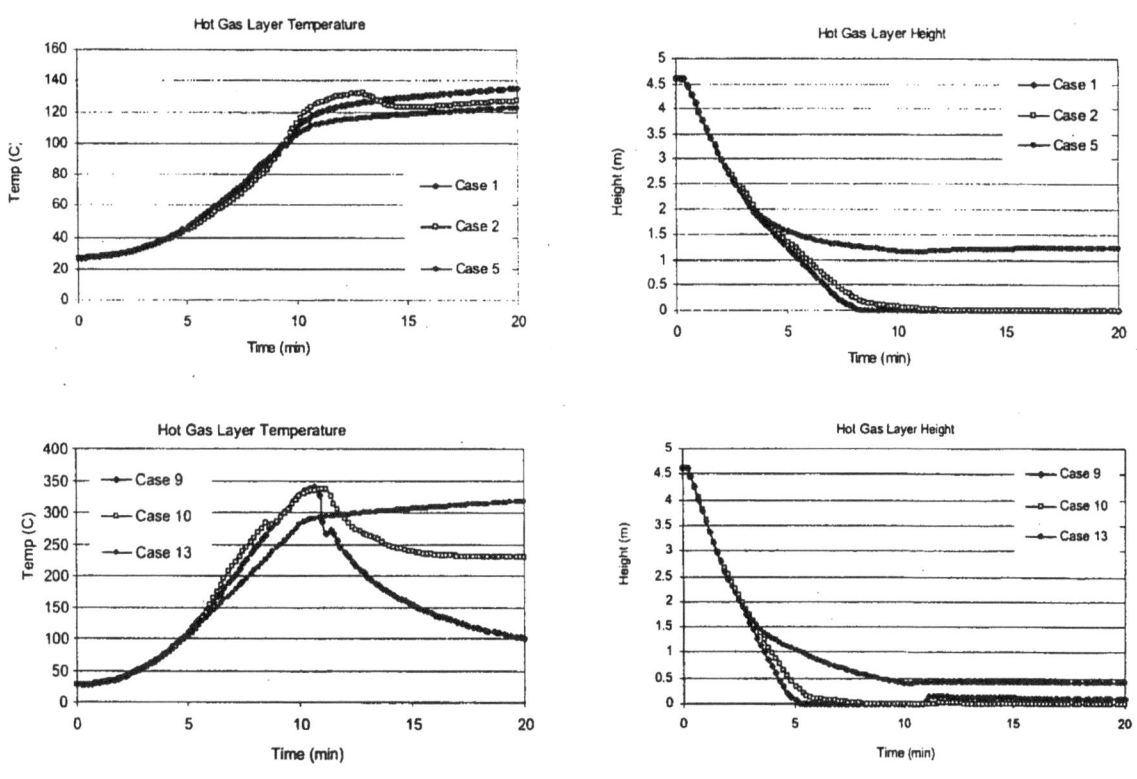

Figure 5-3: Hot Gas Layer Temperature Profiles

5.2.2 Ceiling Jet Temperature

Two sensors were specified in MAGIC's computational domain to record the gas temperature at the specified locations. Figure 5-4 illustrates selected ceiling jet temperature profiles.

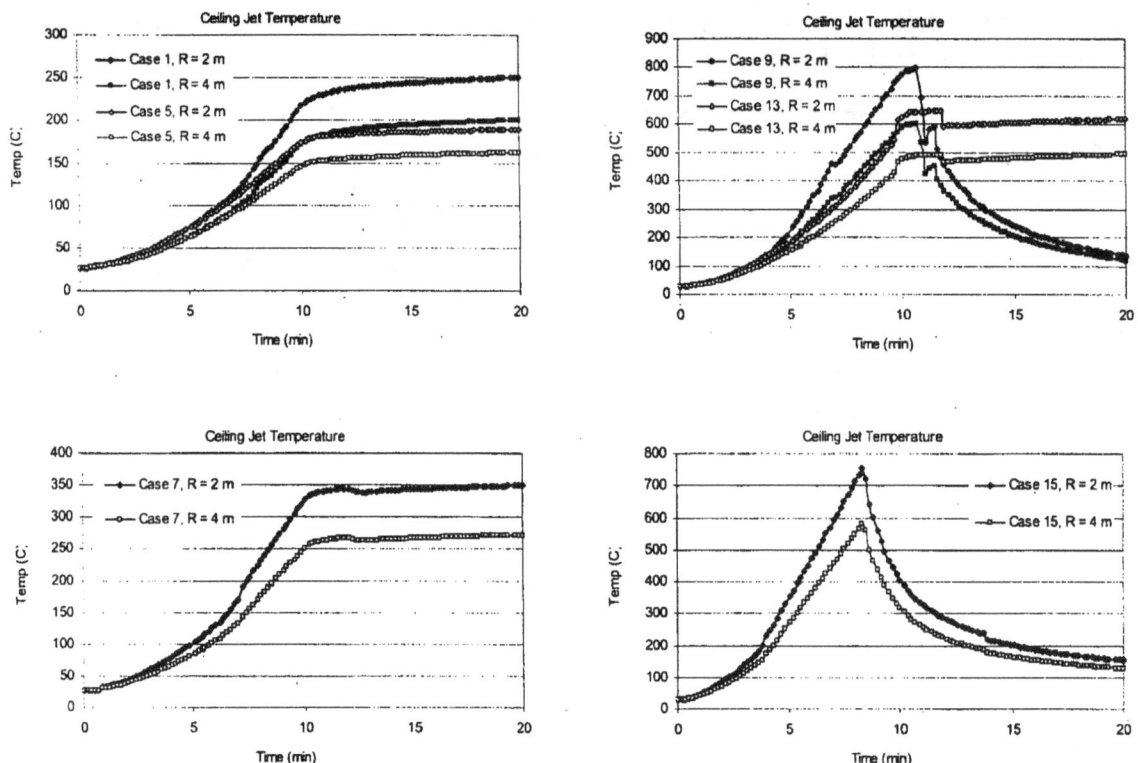

Figure 5-4: Ceiling Jet Temperature Profiles

MAGIC performed as expected. First, for each case, the temperature at a longer radial distance, R, was lower than at a shorter R. Second, the ceiling jet temperature was higher than the predicted hot gas layer temperatures for the respective cases. For example, at a relatively large radial distance from the fire, R = 4 m (13.1 ft), the ceiling jet temperature was just above 150 °C (302 °F) in Case 5. Recall that the predicted hot gas layer temperature for Case 5 was below 150 °C (302 °F). Another interesting observation is that the ceiling jet temperatures are higher in the closed room simulation (Case 1), compared to the open room simulation (Case 5). This behavior was also observed in the corresponding simulations with a 5-MW heat release rate. Consider, for example, the ceiling jet temperature profiles for Case 9. In this case, with an input heat release rate of 5.0 MW, the peak ceiling jet temperatures were above 600 °C (1,112 °F). The decaying nature of the heat release rate profile (resulting from an oxygen-limited environment) is also reflected, similar to the one observed for hot gas layer temperature.

Finally, Cases 7 and 15 are also relevant to the ceiling jet temperature. In this case, the input parameter of interest is the fire elevation (as opposed to the horizontal radial distance and heat release rate). For a fire located 2.3 m (7.6 ft) above the floor, the ceiling jet temperatures were above 250 °C (482 °F). Case 5, which had identical conditions but with a floor base fire, resulted in temperatures more than 50 °C (90 °F) lower. Figure 5-4 illustrates the temperature profiles for the ceiling jet in Cases 7 and 15.

5.2.3 Plume Temperature

Three plume temperature sensors were specified in MAGIC's computational domain, as illustrated in Figure 5-3. The input parameters included in the sensitivity analysis were the elevation of the sensor above the fire and fire intensity. MAGIC again performed as expected. Specifically, plume temperatures were lower as the elevation above the fire increased, temperatures were higher for higher heat release rate profiles, and temperatures were higher than the corresponding hot gas layer temperatures for evaluated cases. Figures 5-5 illustrates the plume temperature profiles for Cases 1 and 9, respectively.

Two important observations can be made regarding Figure 5-5. First, the plume temperature for the lowest sensor ($z = 2.5$ m) in Case 1 reached values above 700 °C (1,292 °F). This is a clear indication that the sensor is just outside the steady-flame region of the fire. In Case 9, however, where the fire intensity was 5.0 MW, all peak plume temperatures were above 1,000 °C (1,832 °F). These values should be interpreted as sensors immersed in flames.

In the case of a fire elevated 2.3 m (7.6 ft) from the floor, MAGIC predicted plume temperatures on the order of thousands of degrees for the lowest two sensors. Peak flame temperatures are generally on the order of 1,500 °C (2,732 °F). This is also an indication of sensors inside the flame.

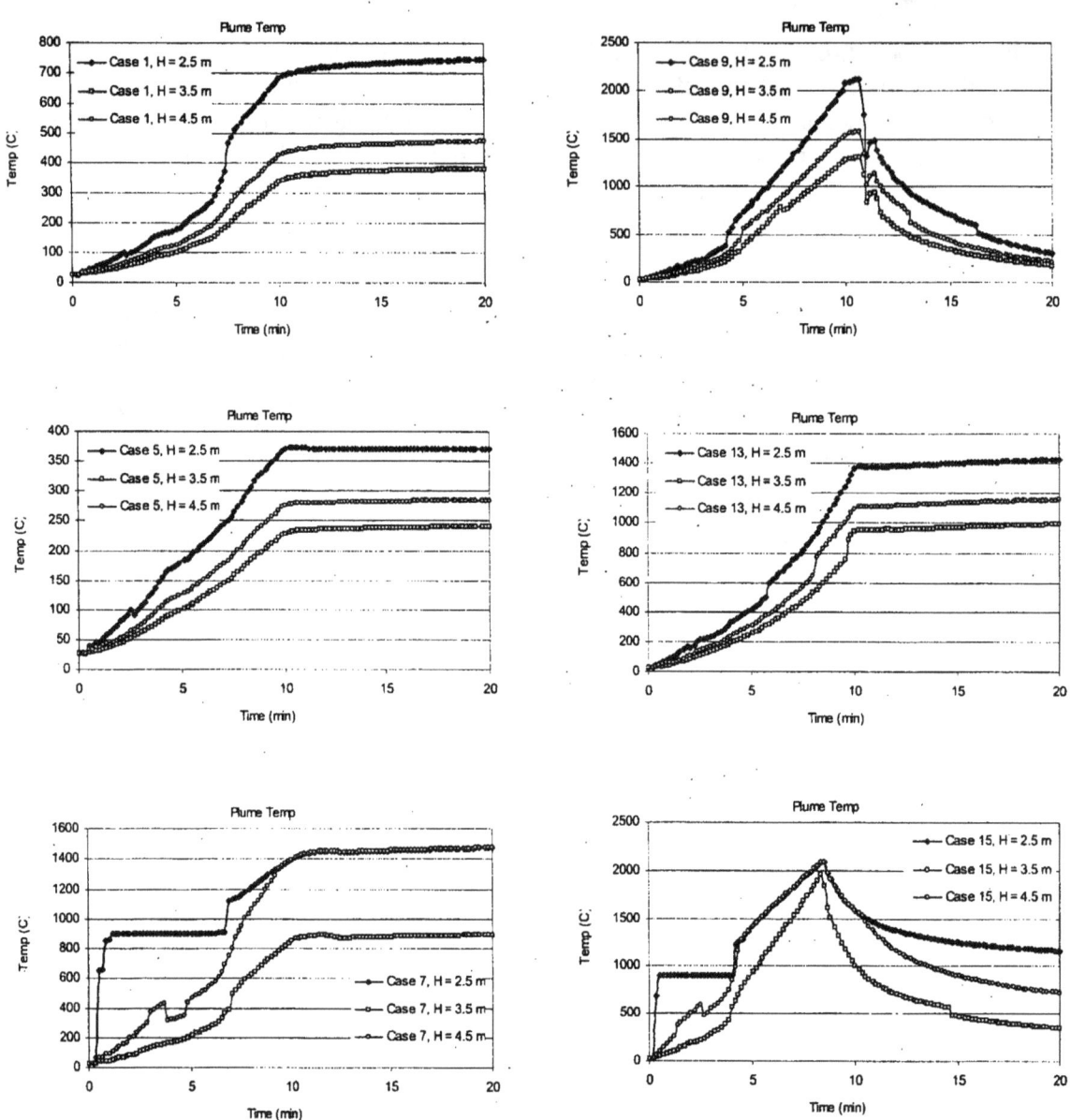

Figure 5-5: Plume Temperature Profiles

5.2.4 Flame Height

The flame height results illustrated in Figure 5-6 suggest two observations. First, the flame presents a linear growth during the t^2 growth period of the fire. Flame height is constant during the steady-burning period. Second, notice that MAGIC predicted flame heights above the ceiling height of 4.6 m (15.2 ft) in Cases 9 and 13. This is a clear indication that flames are reaching the ceiling.

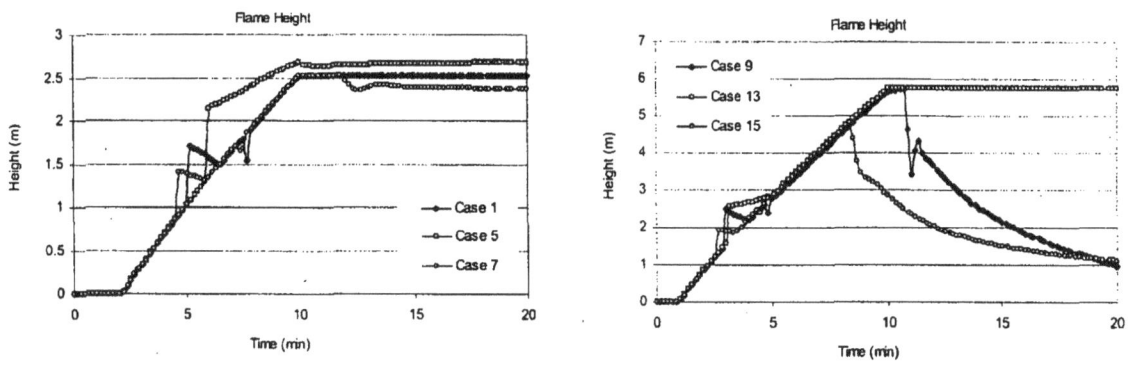

Figure 5-6: Flame Height Profiles

5.2.5 Oxygen Concentration

Two important aspects of modeling oxygen concentrations in commercial NPP scenarios are the amount of oxygen available for combustion in the room and the lower oxygen limit (LOL). These two aspects are, of course, closely related. In terms of the oxygen available for combustion, the fire consumes whatever oxygen is available. As long as there is oxygen above the LOL, the fire will burn at its specified heat release rate. The higher the heat release rate, the greater the amount of the consumed oxygen. Natural and mechanical ventilation conditions will affect the amount of oxygen available. The LOL is a user input, and the most conservative value is 0%. That is, the fire will burn with an intensity governed by the amount of oxygen or fuel until the oxygen in the room has been consumed. Increasing the LOL to a higher value, let's say 10%, will indicate that the fire will not be able to burn if the oxygen concentration is below 10%.

Figure 5-7 illustrates the oxygen concentration profile for Cases 5 and 13, which are simulations with one open door. Notice that the concentration was well above 10 percent. As expected, Case 13 showed a lower oxygen concentration because of the higher heat release rate.

The mechanical ventilation effects in oxygen concentration profiles can be observed in Figure 5-7. Notice that there is more oxygen in Cases 2 and 10, in which the mechanical ventilation system (both injection and extraction) was operating.

The effects of low oxygen concentration on the heat release rate can be observed in simulations with closed doors (specifically Cases 9 and 11). Figure 5-7 compiles the results. Notice that in Case 9, where the LOL is 0%, the heat release rate begins to decay when the oxygen concentration is 0%. In Case 11, where the LOL is 10%, the heat release rate begins to decay at 550 seconds, when the oxygen concentration is 10%. It is interesting to note that the two oxygen concentration profiles are identical up to 10%. At that point, the fire in Case 9 maintains its original intensity and, therefore, consumes more oxygen than the fire in Case 11.

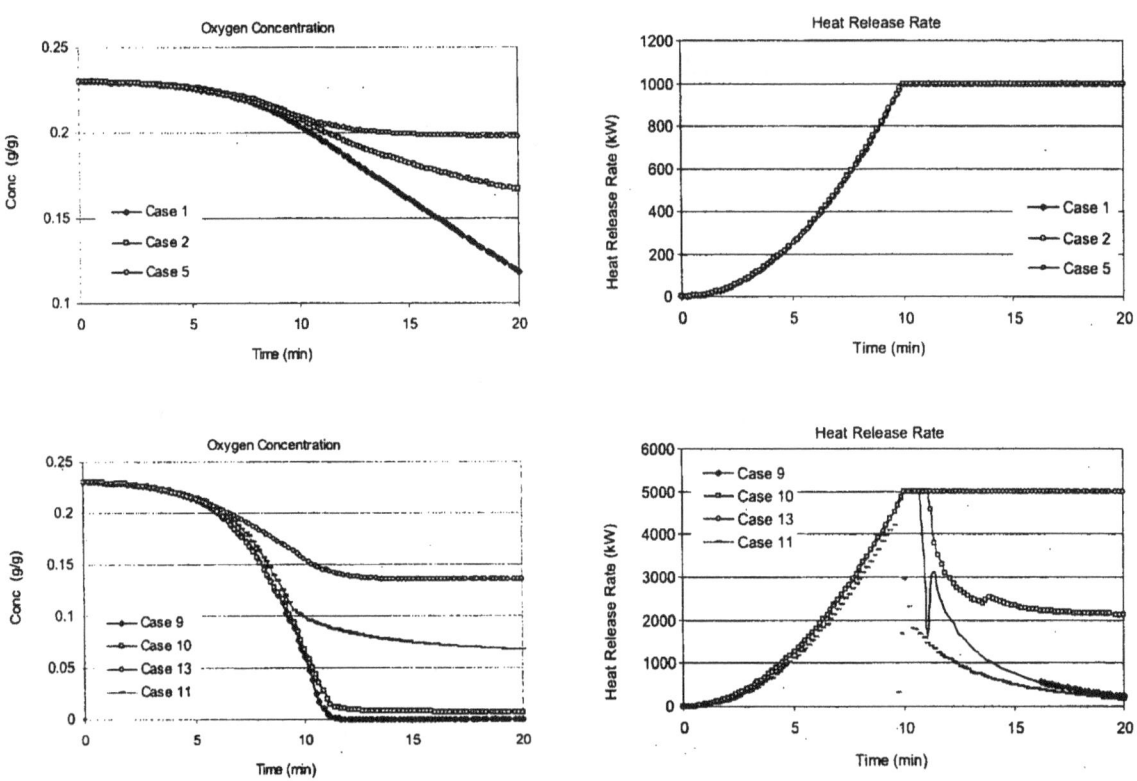

Figure 5-7: Oxygen Concentration Profiles

5.2.6 Smoke Concentration

The relevant output for smoke concentration in MAGIC is the average extinction coefficient, k, with units of 1/m. The relevant input value governing k is the specific area [Ref. 20], s (with units of m^2/g), which is calculated using the soot yield of the fuel, y_s, as follows $s = y_s k_m$, where k_m is a constant value of 7600 m^2/g [Ref. 21]. The average extinction coefficient can be converted to concentration in units of mg/m^3 or visibility in units of m with relatively simple algebraic manipulations. For the purpose of NPP applications, visibility would be the most relevant output. Recall from Reference 21 that the average extinction coefficient correlates linearly with visibility, based on the equation $S = 3/k$ for a light-reflecting object, or $S = 8/k$ for a light-emitting object, where S is the visibility distance in m.

In this sensitivity analysis, Cases 5, 6, 13 and 14 are relevant to visibility. In those cases, the fuel was varied from heptane to toluene in order to explore the effects on the average extinction coefficient. The MAGIC input governing the average extinction coefficient is the specific area, s, which has units of m²/kg. The specific area for heptane is 106.4 m²/kg, while the value for toluene is 1482 m²/kg. Figure 5-8 summarizes the average extinction coefficient results already transformed to units of concentration (mg/m³). The direct output from the model was converted to mg/m³ using the following equation:

$$\upsilon = \frac{k}{k_m}$$

where υ is the concentration in mg/m³, and k_m is a constant with value 0.0076 m²/mg [Ref. 21]. Cases 5 and 6 are associated with a 1.0-MW fire, while Cases 13 and 14 are associated with a 5.0-MW fire. As expected, the highest extinction coefficient resulted from the toluene fuel burning at an intensity of 5.0 MW.

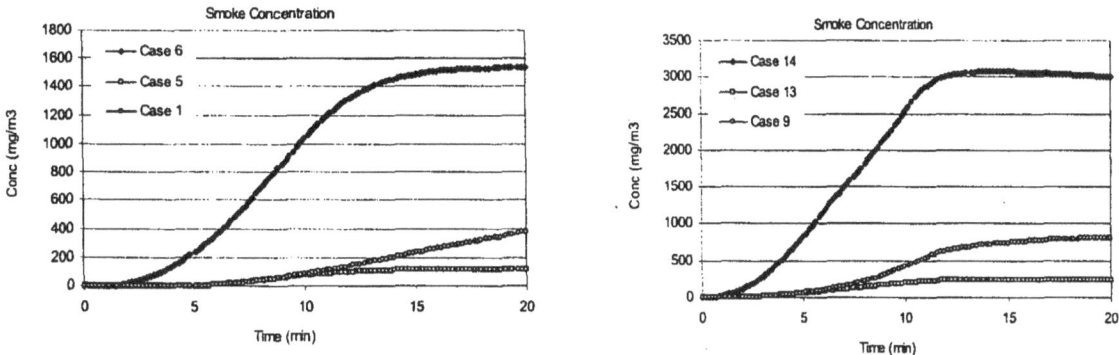

Figure 5-8: Smoke Concentration Profiles

5.2.7 Room Pressure

In addition to the variations in heat release rate, the room openings varied from a leakage path of 0.015 m² (0.16 ft²) to a 5.76-m² (62-ft²) open door in order to explore the impact on room pressure. Figure 5-9 illustrates the pressure profiles for Cases 1, 5, 9, and 13, which were simulations with closed doors (only leakage paths) and open doors for the two heat release rates selected for the study. Given the differences in magnitude, profiles for Cases 5 and 13 should be read on the right y-axis.

In open door simulations (Cases 5 and 13) the pressure at the floor was negative, indicating that fresh air was moving into the enclosure. Recall that the hot gas layer in these simulations did not reach the floor. The region below this hot gas layer interface is associated with the negative pressure profiles in Figure 5-9. In terms of sensitivity to heat release rate, the 5.0-MW fire (Case 13) resulted in higher negative pressure, indicating that air would move into the room at higher velocities than in the case of the 1.0-MW fire. It is interesting to note that the pressurization levels are on the order of Pascals.

By contrast, for rooms with only leakage paths, the pressure profiles were positive (for the most part) and on the order of thousands of Pascals. This is an indication that flows are moving out of the room through the leakage paths. In addition, notice that Case 9 had a negative pressure spike after 600 seconds. As shown in Figure 5-9, this is the time when the heat release rate suddenly decays as a result of an oxygen-limited environment. This pressure spike is attributable to sudden change in heat release rate. At this point, the heat lost to the boundaries is greater than the heat generated by the fire. After this spike, air begins to move into the enclosure through the leakage paths, and the fire is able to burn with an intensity governed by the amount of air drawn into the room. This spike was not observed in Case 1 because the fire had enough oxygen to burn at its specified intensity.

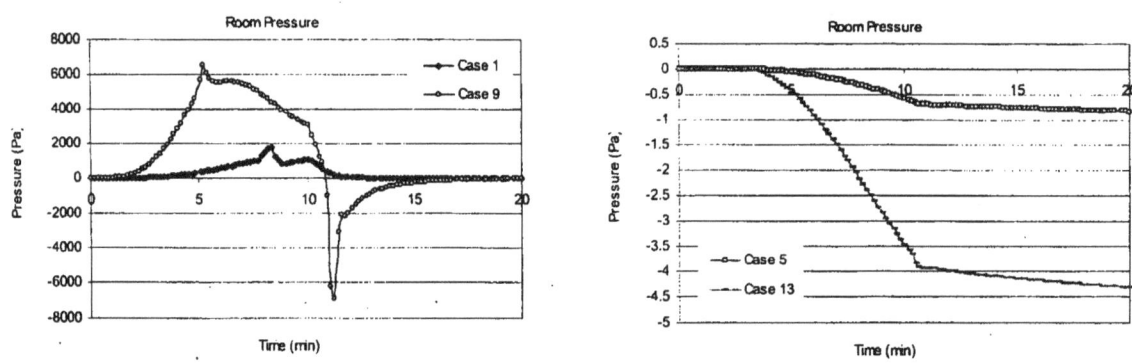

Figure 5-9: Room Pressure Profiles

5.2.8 Target Temperature and Heat Flux

Of primary interest in NPP applications are the effects of the cable's thermo-physical properties on the predicted surface temperature. As illustrated in Figure 5-2, this analysis included two types of cables (XPE and PVC). The two cables have different material properties. The effects of the material properties were explored by comparing surface temperature results for Cases 1 and 9. The only difference between these two cases was the fire heat release rate. As depicted in Figure 5-11, the selected material properties did not have a significant impact on the surface temperature profile. Notice that the profiles are almost identical for the XPE and PVC targets in both cases. However, the damage or ignition temperature was an important distinction.

Another important aspect of evaluating target response in NPP fire scenarios is the difference between gas temperatures at the location of the target and the surface temperature of that target. MAGIC provides both results as part of its output library, and Figure 5-11 illustrates this comparison. In Case 9, the gas temperature was higher than the surface temperature for the first 800 seconds of the simulation, and the highest temperature difference was just above 100 °C. The temperatures then converged when the fire was well into its decay stage. By contrast, the gas temperature was always higher than the surface temperature throughout the simulation in Case 1 and the temperature difference was approximately 50 °C.

MAGIC offers two modeling alternatives for predicting cable temperature, which include modeling the cable as (1) a thermal target or (2) a cable. This section compares the two alternatives. The fundamental difference between the two alternatives is that a cable is treated as a cylinder, while a thermal target is treated as a slab. This shape difference requires the following computational distinctions:

- numerical resolution one-dimensional plane for targets and one-dimensional cylinder for cables
- convective heat exchange coefficient on a plane or cylindrical surface
- configuration factors for the radiative flux calculation

In the following example, a cable is compared to a target. The target is considered similar because the thickness of the target (1) is equal to the radius of the cable, and (2) is calculated conserving the same surface-to-volume ratio. Figure 5-10 illustrates the relationship between the cable and target.

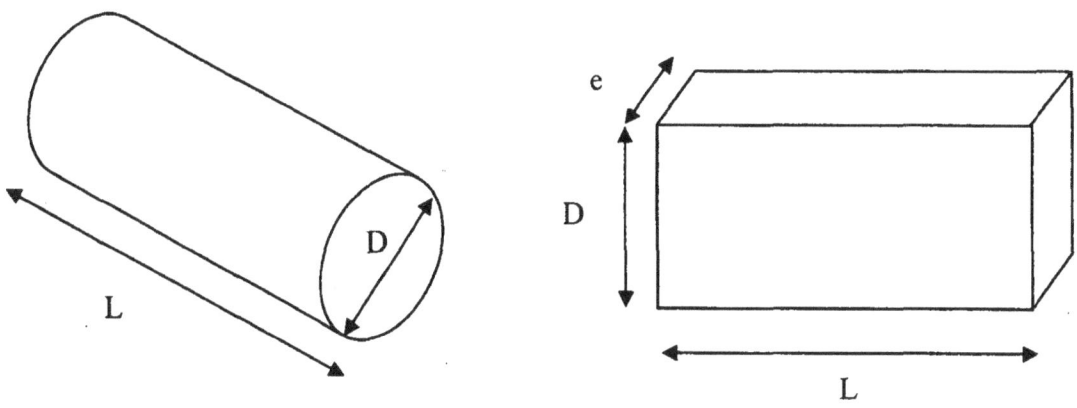

	Cable	Target
Surface Area	πDL	DL
Volume	$\pi \dfrac{D^2}{4} L$	DLe

Figure 5-10: Equivalence between Cable and Targets

The surface-to-volume ratio gives a thickness target value of $e = \dfrac{D}{4}$. Figure 5-11 includes temperature profiles with targets with thicknesses D/4 and D/2. Targets with thickness D/4 resulted with the highest surface temperature.

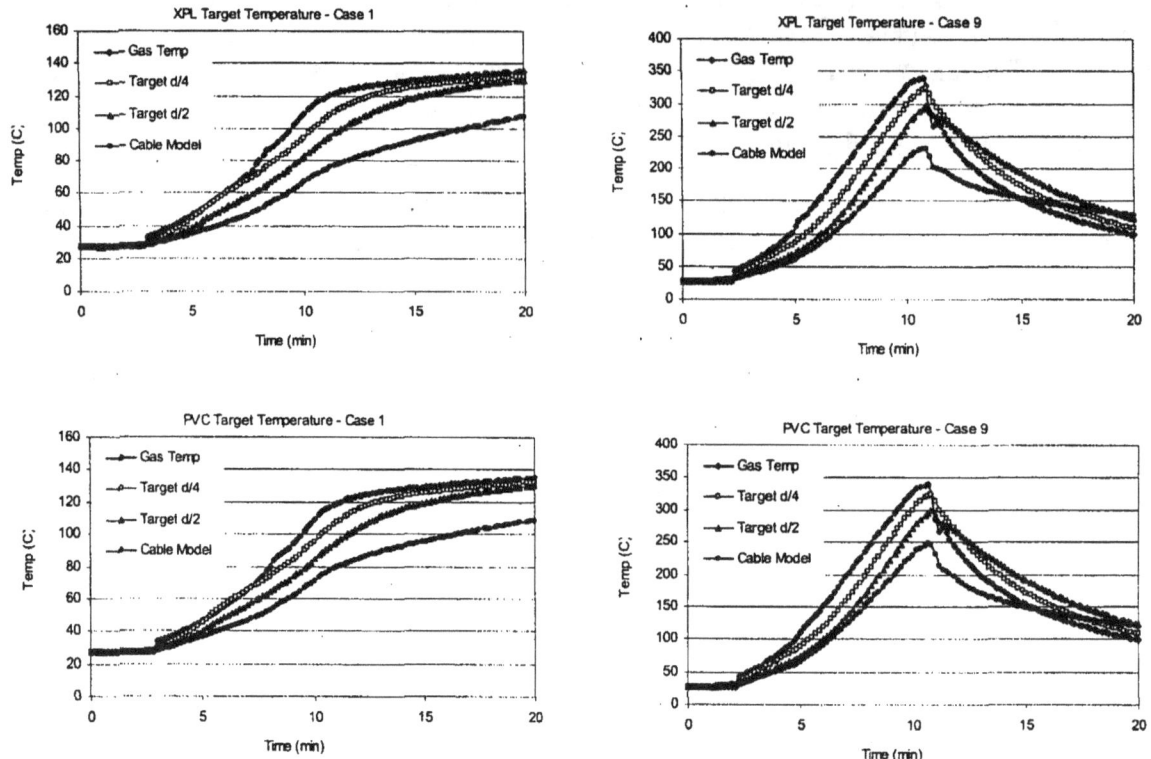

Figure 5-11: Target Temperature Profiles

In terms of heat flux, MAGIC's "Total Heat Flux" and "Incident Heat Flux" output options are relevant in this study. The former is the total radiative and convective heat flux contributions to the target. The latter is total radiative heat flux received by the target. That is, the incident flux is the radiation flux received from the environment. It is normally positive because it is not a balance. At the beginning of the simulation conditions are ambient. Therefore, the incident heat flux is approximately $\varepsilon \sigma T_{amb}^4 \approx 0.9(5.67E-11)(293)^4 \approx 0.4\,kW/m^2$. The thickness of the target does not affect these output options.

As illustrated in Figure 5-2, PVC and XPE targets were located 3.55 m (11.65 ft) away from the fire, and the elevation of both targets was 2.3 m (7.6 ft). In MAGIC, these targets serve as sensors and record thermal conditions in their specified location. Each case study exhibited the identical predicted heat flux to each target type. That is, given the symmetrical arrangement of targets relative to the fire source, both XPE and PVC targets receive the same radiated heat flux in each case. As expected, the total heat flux is higher than the incident heat flux attributable to the contribution of the convective heat transfer.

Finally, the fuel type appears to have some effect on thermal radiation levels. According to the results, the simulations conducted with heptane fires produced higher heat fluxes than the corresponding simulations conducted with toluene fires (Cases 6 & 14), although the magnitude differences were less than 1 kW/m². Figures 5-12 and 5-13 compile the graphical results.

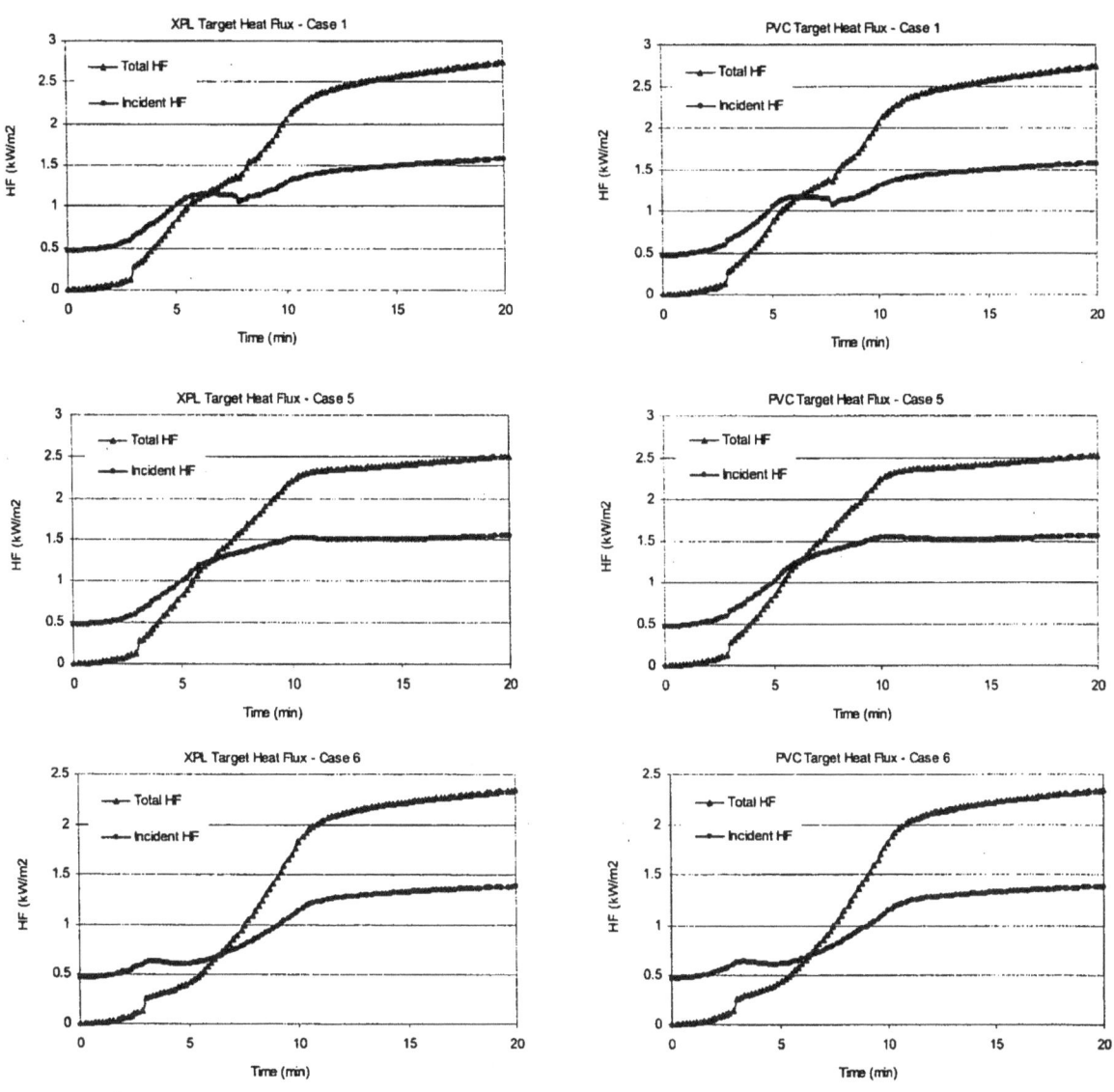

Figure 5-12: Target Heat Flux Profiles, 1-MW Fire

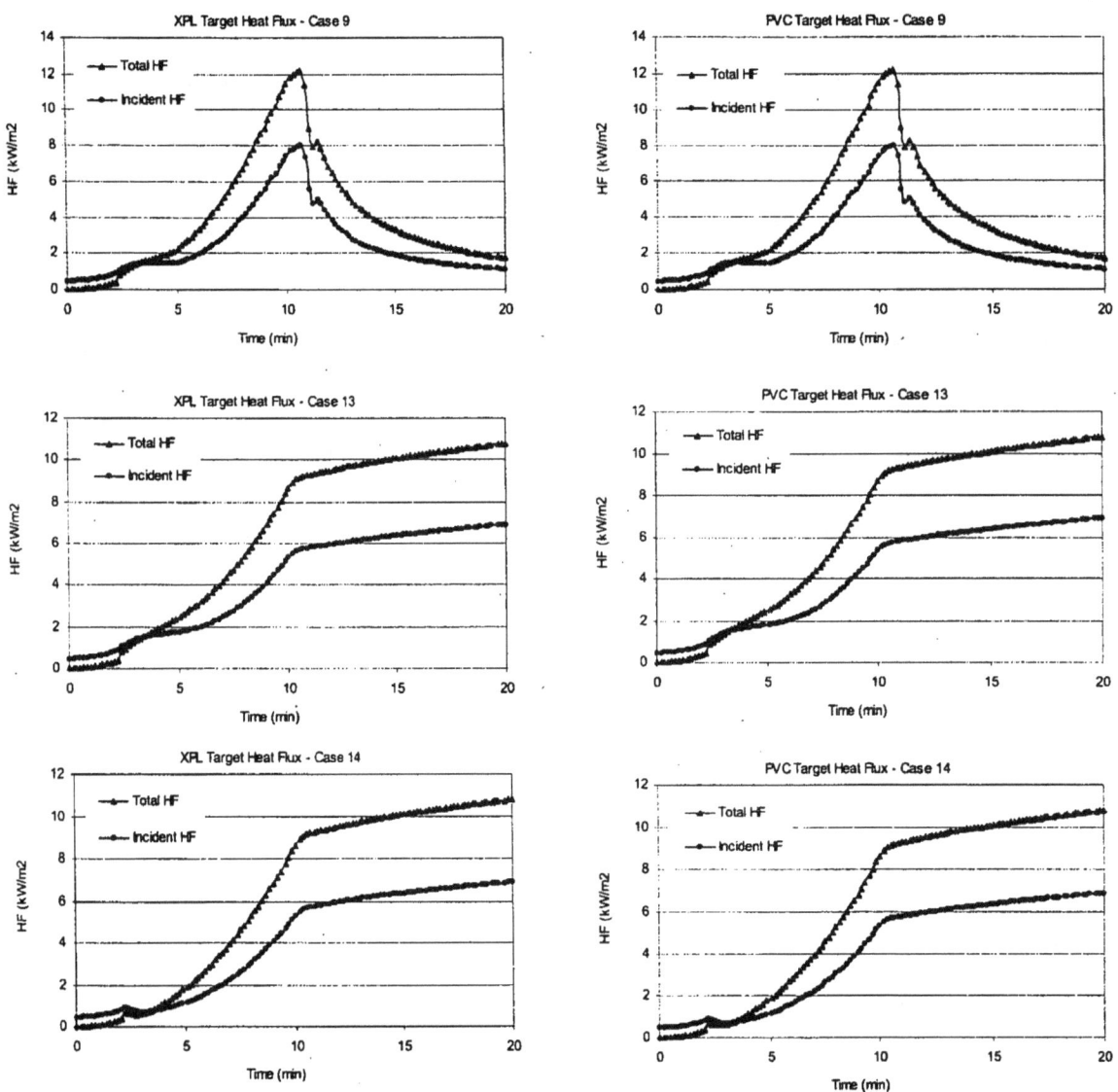

Figure 5-13: Target Heat Flux Profiles, 5-MW Fire

5.2.9 Wall Temperature

Figures 5-14 and 5-15 summarize a comparison between the use of MAGIC's "Wall Temperature" and "Surface Temperature of the Target" output options. The "Wall Temperature" output results from a one-dimensional finite difference calculation of conduction into the walls. The internal boundary conditions are the thermal properties of the gas in contact with the particular wall surface (upper or lower gas layer). The "Surface Temperature of the Target" output option results from a calculation of conduction into a slab with a thickness similar to that of the wall.

This comparison is important because the validation study described in Chapter 6 for wall temperature was developed using the latter option. That is, virtual sensors in MAGIC (e.g., targets) were specified in the same location as the wall thermocouples in the experimental series. The targets were specified with the same thermo-physical properties and thickness as the walls. The only difference in the specification is the emissivity. The targets had an emmisivity of 0.95 and MAGIC does not require emissivity as a wall property input.

Results suggests that the use of the "Target" feature in MAGIC for predicting wall surface temperature can produce higher temperatures, with the exception of the floor surface in open-door tests. In the cases run with a 1-MW fire, the temperature difference between both modeling strategies is approximately 10 °C. In cases run with a 5-MW fire, the temperature difference is between 40 °C and 60 °C.

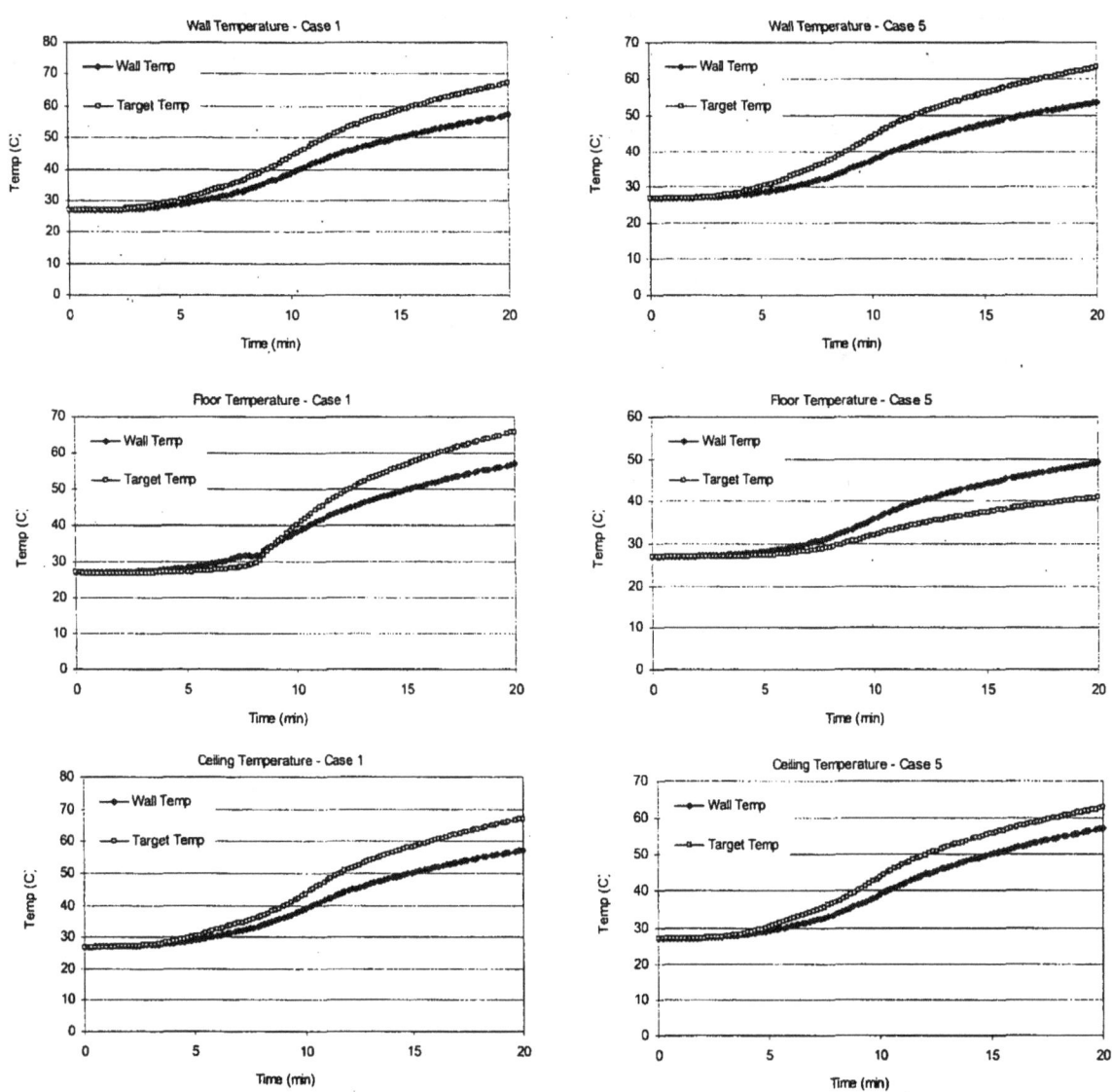

Figure 5-14: Target Temperature vs. Wall Temperature, 1-MW Fire

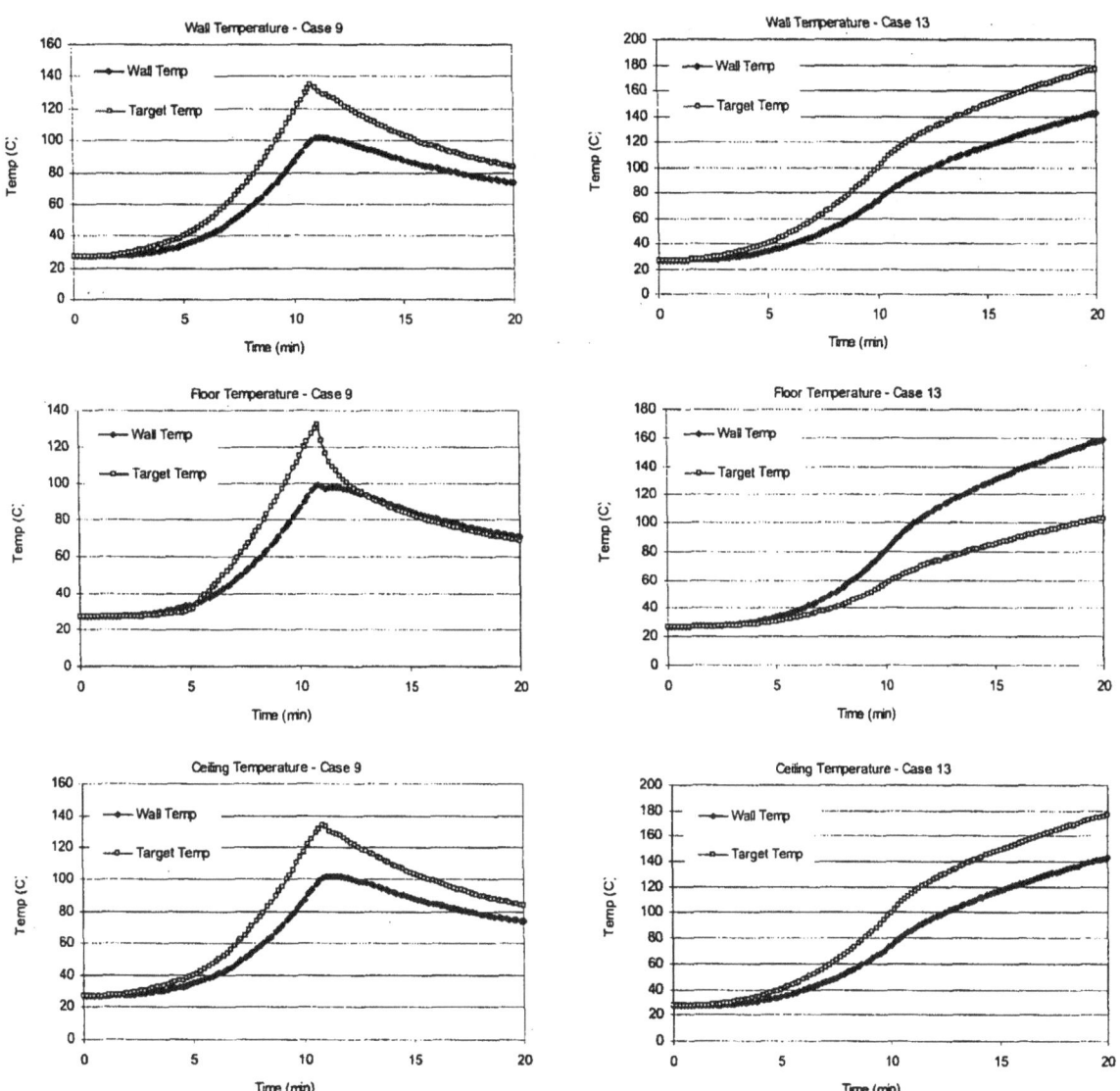

Figure 5-15: Target Temperature vs. Wall Temperature, 5-MW Fire

5.3 Concluding Remarks

This chapter illustrates the effect of the most important parameters in fire modeling with the MAGIC code. The set of models included in MAGIC are intended to translate the impact of those parameters on the fire-generated conditions in a compartment. Therefore, it is important to understand the effects of the input parameters on the predicted fire conditions, considering that the simulation results are simplifications and idealizations of real fire-induced temperatures and flows.

It is difficult to generalize which input parameters are more important than others because it depends on specific applications, and most (if not all) of the parameters are mathematically related. As illustrated in this chapter, different parameters are important for different sub-models. In most applications, the fire modeling analyst will need to determine which outputs are relevant for the scenario under evaluation, which parameters will affect those outputs, and how variations in those parameters will impact the conclusions made from the simulation results.

6
MODEL VALIDATION

This chapter summarizes the results of a validation study conducted for the zone model MAGIC, in which its predictions are compared with measurements collected from six sets of large-scale fire experiments. A brief description of each set of experiments is given here. Further details can be found in Volume 2 and in the individual test reports.

ICFMP BE #2: Benchmark Exercise #2 consists of eight experiments, representing three sets of conditions, to study the movement of smoke in a large hall with a sloped ceiling. The results of the experiments were contributed to the International Collaborative Fire Model Project (ICFMP) for use in evaluating model predictions of fires in larger volumes representative of turbine halls in NPPs. The tests were conducted inside the VTT Fire Test Hall, which has dimensions of 19 m high x 27 m long x 14 m wide (62 ft x 88.5 ft x 46 ft). Each case involved a single heptane pool fire, ranging from 2 MW to 4 MW.

ICFMP BE #3: Benchmark Exercise #3, conducted as part of the ICFMP and sponsored by the NRC, consists of 15 large-scale tests performed at NIST in June 2003. The fire sizes range from 350 kW to 2.2 MW in a compartment with dimensions of 21.7 m x 7.1 m x 3.8 m (71 ft x 23 ft x 12.5 ft), designed to represent a variety of spaces in a NPP containing power and control cables. The walls and ceiling are covered with two layers of marinate boards, while the floor is covered with two layers of gypsum boards. The room has one door with dimensions of 2 m x 2 m (6.6 ft x 6.6 ft), and a mechanical air injection and extraction system. Ventilation conditions and fire size and location are varied, and the numerous experimental measurements include gas and surface temperatures, heat fluxes, and gas velocities.

ICFMP BE #4: Benchmark Exercise #4 consists of kerosene pool fire experiments conducted at the Institut für Baustoffe, Massivbau und Brandschutz (iBMB) of the Braunschweig University of Technology in Germany. The results of two experiments were contributed to the ICFMP. These fire experiments involve relatively large fires in a relatively small [3.6 m x 3.6 m x 5.7 m (12 ft x 12 ft x 19 ft)] concrete enclosure. Only one of the two experiments (Test 1) was selected for the present V&V study.

ICFMP BE #5: Benchmark Exercise #5 consists of fire experiments conducted with realistically routed cable trays in the same test compartment as BE #4. The compartment was configured slightly differently, and the height was 5.6 m (18.4 ft) in BE #5. Only Test 4 was selected for the present evaluation, and only the first 20 minutes, during which an ethanol pool fire preheated the compartment.

<u>FM/SNL Series</u>: The Factory Mutual & Sandia National Laboratories (FM/SNL) Test Series is a series of 25 fire tests conducted for the NRC by Factory Mutual Research Corporation (FMRC), under the direction of Sandia National Laboratories (SNL). The primary purpose of these tests was to provide data with which to validate computer models for various types of NPP compartments. The experiments were conducted in an enclosure measuring 60 ft long x 40 ft wide x 20 ft high (18 m x 12 m x 6 m), constructed at the FMRC fire test facility in Rhode Island. All of the tests involved forced ventilation to simulate typical NPP installation practices. The fires consist of a simple gas burner, a heptane pool, a methanol pool, or a polymethyl-methacrylate (PMMA) solid fire. Four of these tests were conducted with a full-scale control room mockup in place. Parameters varied during testing are the heat release rate, enclosure ventilation rate, and fire location. Only three of these tests (Tests 4, 5 and 21) were used in the present evaluation. Test 21 involved the full-scale mockup. All were gas burner fires.

<u>NBS Multi-Room Series</u>: The National Bureau of Standards (NBS, now the National Institute of Standards and Technology, NIST) Multi-Compartment Test Series consists of 45 fire tests representing 9 different sets of conditions, with multiple replicates of each set, which were conducted in a three-room suite. The suite consists of two relatively small rooms, connected via a relatively long corridor. The fire source, a gas burner, is located against the rear wall of one of the small compartments. Fire tests of 100, 300, and 500 kW were conducted, but only three 100-kW fire experiments (Test 100A, 100O, and 100Z) were used for the current V&V study.

Technical details of the calculations, including output of the model and comparison with experimental data are provided in Appendix A. The results are organized by quantity as follows:

- Section 6.1: Hot Gas Layer Temperature And Height
- Section 6.2: Ceiling Jet Temperature
- Section 6.3: Plume Temperature
- Section 6.4: Flame Height
- Section 6.5: Oxygen Concentration
- Section 6.6: Smoke Concentration
- Section 6.7: Compartment Pressure
- Section 6.8: Target Temperature and Heat Flux
- Section 6.9: Wall Heat Flux and Temperature

The model predictions are compared to the experimental measurements in terms of the relative difference between the maximum (or where appropriate, minimum) values of each time history:

$$\varepsilon = \frac{\Delta M - \Delta E}{\Delta E} = \frac{\left(M_p - M_o\right) - \left(E_p - E_o\right)}{\left(E_p - E_o\right)}$$

ΔM is the difference between the peak value of the model prediction, M_p, and its original value, M_o. ΔE is the difference between the experimental measurement, E_p, and its original value, E_o.

A positive value of the relative difference indicates that the model over-predicted the severity of the fire (e.g., a higher temperature, lower oxygen concentration, higher smoke concentration, etc.).

Each section in this chapter contains a scatter plot that summarizes the relative difference results for all of the predictions and measurements of the quantity under consideration. The details of the calculations, input assumptions, and time histories of the predicted and measured output are included in Appendix A. Only a brief discussion of the results is included in this chapter. At the end of each section, a color rating is assigned to each of the output category, indicating, in a very broad sense, how well the model treats that particular quantity. Colors are assigned based on the following criteria. Once the user determines that the validation results reported here are applicable (see Volume 1), the user must determine the predictive capability of the fire models. The following two criteria are used to characterize the predictive capability of the model:

Criterion 1: *Are the physics of the model appropriate for the calculation being made?* This criterion reflects an evaluation of the underlying physics described by the model and the physics of the fire scenario. Generally, the scope of this study is limited to the fire scenarios that are within the stated capability of the selected fire models (e.g., this study does not address the fire scenarios that involve flame spread within single and multiple cable trays).

Criterion 2: *Are there calculated relative differences outside the experimental and model input uncertainty?* This criterion is used as an indication of the accuracy of the model prediction. Because fire experiments are used as a way of establishing confidence in model prediction, the confidence can only be as good as our experiments and the model inputs derived from experiments. Therefore, if model predictions fall within the ranges of these combined uncertainties, the predictions are determined to be as accurate as the experiments and data. Section 2.6.3 and Volume 2 of this report provide an introduction and technical details for the uncertainty analysis.

The predictive capability of the model is characterized as follows based on the above criteria:

GREEN: If both criteria are satisfied (i.e., the model physics are appropriate for the calculation being made, and the calculated relative differences are within or very near experimental uncertainty), the V&V team concludes that the fire model prediction is accurate for the ranges of experiments in this study, as described in Tables 2-4 and 2-5 in Volume 1 of this report. A grade of Green indicates that the model can be used with confidence to calculate the specific attribute. The user should recognize, however, that the accuracy of the model prediction is still somewhat uncertain and, for some attributes (such as smoke concentration and room pressure), these uncertainties may be rather large. It is important to note that a grade of Green indicates validation only in the parameter space defined by the test series used in this study. That is, the model is validated when it is used within the ranges of the parameters defined by the experiments.

YELLOW±: If the first criterion is satisfied, and the calculated relative differences are outside the experimental uncertainty but indicate a consistent pattern of model over-prediction or under-prediction, the model's predictive capability is characterized as Yellow+ for over-prediction, and Yellow– for under-prediction. The model prediction for the specific attribute may be useful within the ranges of experiments in this study, and as described in Tables 2-4 and 2-5 in Volume 1, but users should use caution when interpreting the model's results. A complete understanding of model assumptions and scenario applicability to these V&V results is necessary. The model may be used if the grade is Yellow+ when the user ensures that model over-prediction reflects conservatism. The user must exercise caution when using models with capabilities characterized as Yellow±.

`YELLOW`: If the first criterion is satisfied, and the calculated relative differences are outside experimental uncertainty with no consistent pattern of over- or under-prediction, the model's predictive capability is characterized as Yellow. A Yellow classification is also used despite a consistent pattern of under- or over-prediction if the experimental data set is limited. Caution should be exercised when using a fire model for predicting these attributes. In this case, users are referred to the details of the experimental conditions and validation results documented in Volumes 2 through 6. Users are advised to review and understand the model assumptions and inputs, as well as the conditions and results, to determine and justify the appropriateness of the model's prediction for the fire scenario for which it is being used.

▮: If the first criterion is not met, the particular fire model's capability should not be used.

No color: This V&V study did not investigate this capability. This may be attributable to one or more factors, including unavailability of appropriate data or lack of model, sub-model, or output.

As suggested in the criteria above, there is a level of engineering judgment in the classification of fire model predictive capabilities. Specifically, the V&V project team exercised engineering judgment in the following two areas:

1. Evaluation of the modeling capabilities of the particular tool if the model physics are appropriate.

2. Evaluation of the magnitude of relative differences when compared to the experimental uncertainty. Judgment in this area impacts the determination of Green versus Yellow color.

The team included fire model developers, nuclear power plant fire modeling experts, and code users. In general, a Green or Yellow classification suggests that the V&V team determined that the model physics are appropriate for the calculation been made, within the assumptions of the specific model. The difference between the colors is attributable to the magnitude of the calculated relative differences. Judgment considerations include general experimental conditions, experimental data quality, and the characterization of the experimental uncertainty.

6.1 Hot Gas Layer Temperature and Height

The single most important prediction a fire model can make is the temperature of the hot gas layer. After all, the impact of the fire is often assessed not only a function of the heat release rate, but also as a function of the compartment temperature. A good prediction of the height of the hot gas layer is largely a consequence of a good prediction of its temperature because smoke and heat are largely transported together, and most numerical models describe the transport of both with the same type of algorithm. The following is a summary of the accuracy assessment for the hot gas layer predictions of the six test series:

ICFMP BE #2: MAGIC under-predicts the hot gas layer temperature by less than 10% for all three cases. This falls within the range of experimental uncertainty. In addition, MAGIC under-predicts the hot gas layer height by less than 10% in all three cases. That is, MAGIC's height prediction is above the measured height. A graphical comparison of the MAGIC predictions and experimental observations for these three cases is presented in Figure A-2. The scatter plot in Figure 6-1 illustrates the relative differences between the measured and predicted peak hot gas layer temperatures and heights.

ICFMP BE #3: MAGIC predicts the hot gas layer temperature and height to within experimental uncertainty for all 15 tests. It should be noted that the discrepancies in the hot gas layer heights depicted in Figures A-4 and A-5 (which refers to closed-door tests) are attributable to the data reduction method used to determine the experimental layer interface. This method is not applicable for tests in which a single gas layer develops (i.e., the room is completely filled with smoke). Notice that MAGIC predicts that the hot gas layer eventually reaches the floor, generating a single gas layer in the room. That prediction is consistent with visual observations during the experiments. Given the inconsistency between model results and the reduced experimental data, no relative differences were calculated for closed-door tests.

The collection of graphical comparisons between MAGIC predictions for hot gas layer temperatures and heights for ICFMP BE #3 is presented in Figures A-4 through A-7. The relative differences calculated for peak values are summarized in Table A-2 and Figure 6-1.

ICFMP BE #4: MAGIC predicts the hot gas layer temperature within experimental uncertainty for the single test (Test 1). However, there is some discrepancy in the shapes of the curves for the hot gas layer height (see Figure A-11). This discrepancy is associated with a relative difference of 25%, which is outside the range of experimental uncertainty. A possible explanation for the discrepancy in the layer height is the fact that the room was almost engulfed in flames, which may not be consistent with the fundamental assumption in MAGIC of two distinct gas layers. The relative differences for layer temperature and height are also plotted in Figure 6-1.

ICFMP BE #5: MAGIC predicts the hot gas layer temperature and height to within experimental uncertainty for the single test (Test 4). The graphical comparison between experimental measurements and model predictions, illustrated in Figure A-11 suggests very good agreement between the profiles. The calculated relative differences for peak hot gas layer temperature and height are listed in Table A-4.

FM/SNL: MAGIC predicts the hot gas layer temperature to within experimental uncertainty for Tests 4, 5, and 21. In the case of the hot gas layer height, there are inconsistencies in the comparison of hot experimental measurements and model predictions. As previously discussed for the case of closed-door tests in ICFMP BE #3, the data reduction method for determining hot gas layer height is not applicable for closed-door tests. Consequently, the graphical comparisons presented in Figure A-13 do not show good agreement between model predictions and experimental measurements. For that same reason, no relative differences were calculated for hot gas layer height in this test series.

NBS Multi-Room: MAGIC's hot gas layer temperature predictions in this test series are, for the most part, outside the experimental uncertainty for both the fire room and adjacent compartments. At the same time, the comparisons between predicted and observed values suggest over-predictions of hot gas layer temperatures, as depicted in Figures A-15 through A-17 and Figure 6-1. The hot gas layer height predictions were close to the experimental uncertainty limits with some over-predictions in the corridor. Recall that a negative relative difference suggests a model hot gas layer height prediction above the experimental observation.

Summary: HGL Temperature and Height (Green for single rooms and Yellow+ for multiple rooms)

The hot gas layer temperature and height merit a Green classification for single rooms and a Yellow+ classification for scenarios with multiple rooms.

- The research team considers the MAGIC model for calculating hot gas layer temperatures to be appropriate for its intended applications.

- MAGIC's predictions of the hot gas layer temperature and height are, with the exception of the selected tests from the NBS test series, within experimental uncertainty of 13%.

- The scatter plot in Figure 6-1 summarizes the relative differences calculated for hot gas layer temperatures and height. As previously explained, no relative differences were calculated for closed-door tests.

- Validation results suggest that MAGIC is certainly suited for predicting hot gas layer temperatures and heights in scenarios where this study is applicable. Because most of the validation results are within experimental uncertainty, and MAGIC is over-predicting hot gas layer temperatures in the selected tests from the NBS test series, a color assignment of Green is assigned for single-room fire scenarios and Yellow+ for scenarios with multiple rooms. In the case of hot gas layer height, a Green classification is assigned for the room of fire origin and Yellow for adjacent rooms.

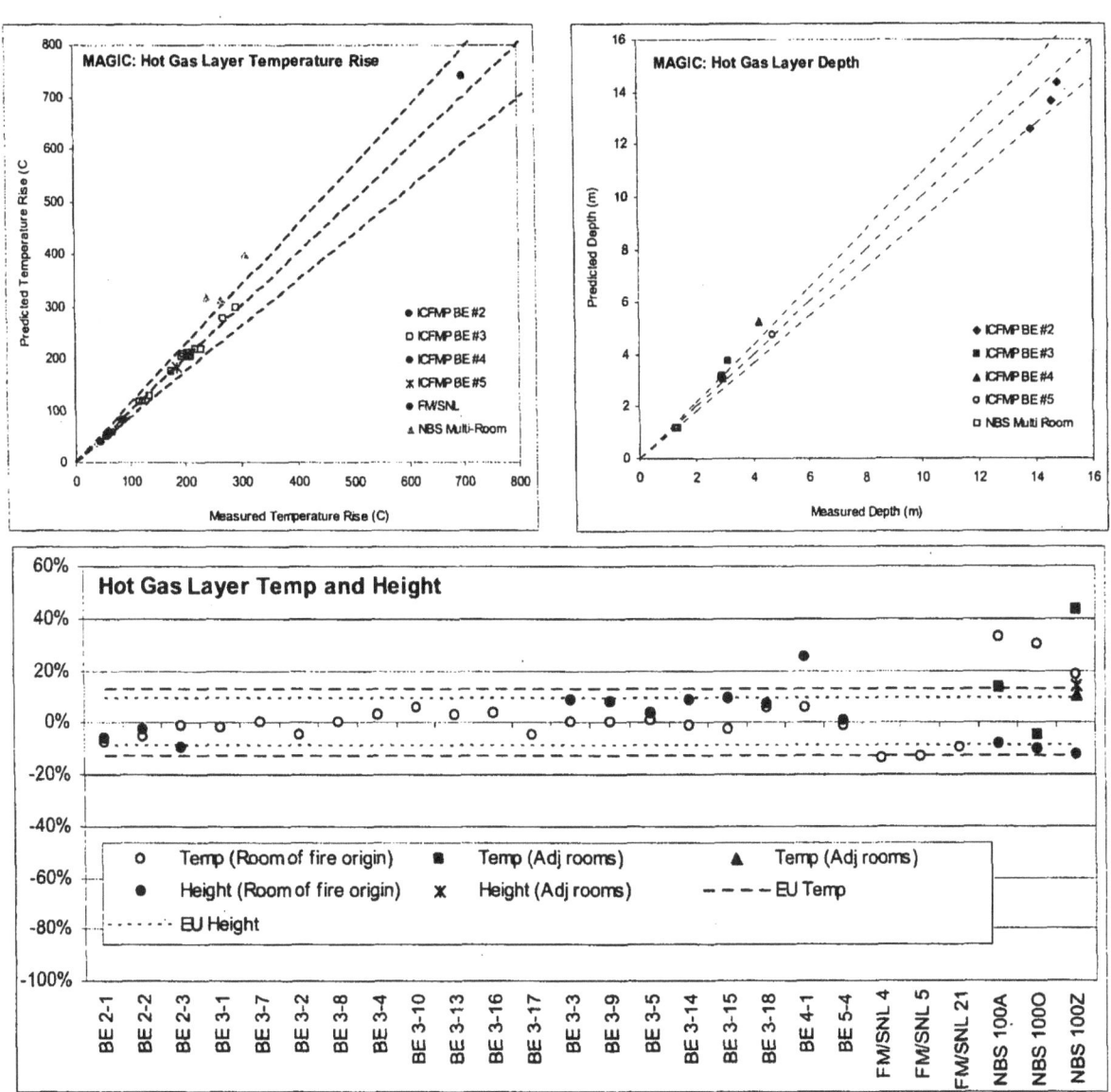

Figure 6-1: Scatter Plot of Relative Differences for Hot Gas Layer Temperature and Height in ICFMP BE #2, #3, #4, and #5, and the Selected FM/SNL and NBS Tests [Experimental uncertainties (EU) are 13% (HGL temp) and 9% (HGL height).]

6.2 Ceiling Jet Temperature

The ceiling jet algorithm in MAGIC consists primarily of the model proposed by Cooper [Ref. 7]. It is recommended that analysts review MAGIC's technical reference [Ref. 2] for specific details about the implementation of the ceiling jet algorithm. Typical of ceiling jet correlations, MAGIC's algorithm applies only to the flow of hot gases under a flat ceiling. Only two of the six test series (ICFMP BE #3 and FM/SNL) involved a ceiling jet formed over a relatively wide, flat ceiling.

ICFMP BE #3: MAGIC predicts the ceiling jet temperature to within experimental uncertainty (16%), with the exception of Tests 10, 13, and 16, as illustrated in Figure 6-2. Interestingly, Test 10 is a replicate of Test 4, which was predicted to within experimental uncertainty. The ventilation system was on during these two tests, and the inconsistent results may be attributed to it. It is difficult to draw conclusions about over-predictions in Tests 15 and 16. The only difference in Tests 13 and 16 is the mechanical ventilation, which was off in Test 13 and on in Test 16. The over-prediction for Test 16 does not appear to be a consistent pattern from MAGIC in predicting ceiling jet temperatures, in particular, because such over-prediction was not observed in similar test in ICFMP BE #3. Therefore, such over-prediction is not considered dominant in assessing MAGIC's capabilities for predicting ceiling jet temperature. Figure 6-2 also suggests that the relative differences for open-door tests are smaller (near 0%) than those in closed-door tests. Furthermore, only two under-predictions, –7% in Test 1 and –1% in Test 14 were calculated.

The graphical comparisons between experimental measurements and MAGIC's predictions for ceiling jet temperature are grouped in Figures A-18 and A-19. Table A-9 lists the calculated relative differences.

FM/SNL: MAGIC predicts the ceiling jet temperature at two locations in Test 4, 5, and 21 to within experimental uncertainty. The graphical comparisons are provided in Figure A-20. The calculated relative differences are listed in Table A-10 and plotted in Figure 6-2.

<u>**Summary**</u>: **Ceiling Jet Temperature — Green**

- With three exceptions corresponding to closed-door tests in ICFMP BE #3, MAGIC's ceiling jet predictions are within experimental uncertainty.

- MAGIC's ceiling jet sub-model is well-suited for the range of scenarios validated in this study. In general, this validation applies to ceiling jet flows under flat unobstructed ceilings and an r/H up to 1.7 (r is the horizontal radial distance and H is the distance between the fire source and the ceiling). Notice that the MAGIC technical manual (Ref. 2) suggests that "the model is valid up to r/H = 3. In MAGIC, they are applied up to r/H =10, to avoid a discontinuity in gas temperature. In cases where the ceiling jet exceeds r/H = 10, it is assumed that the gas temperature beyond r/H = 10 equals the temperature at r/H = 10. These hypotheses are conservative even if they have not been explored experimentally."

- This V&V study included the evaluation of MAGIC's capabilities for predicting ceiling jet temperature inside the hot gas layer.

- Based on the model's robustness and the fact that most relative differences are within experimental uncertainty (16%), a Green classification is assigned.

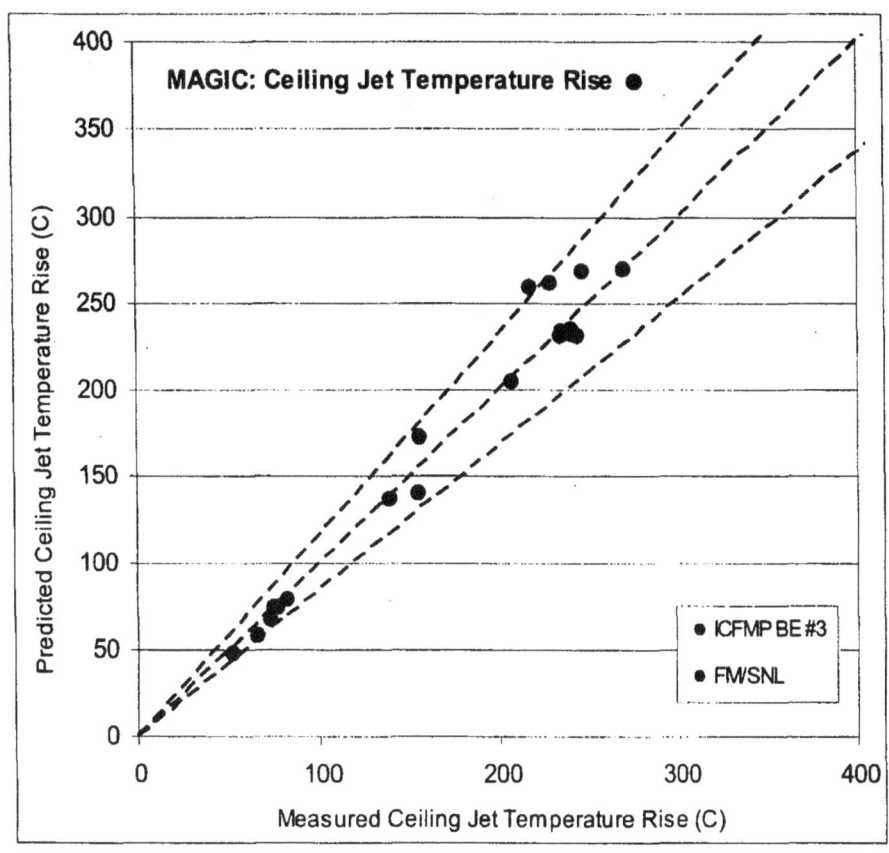

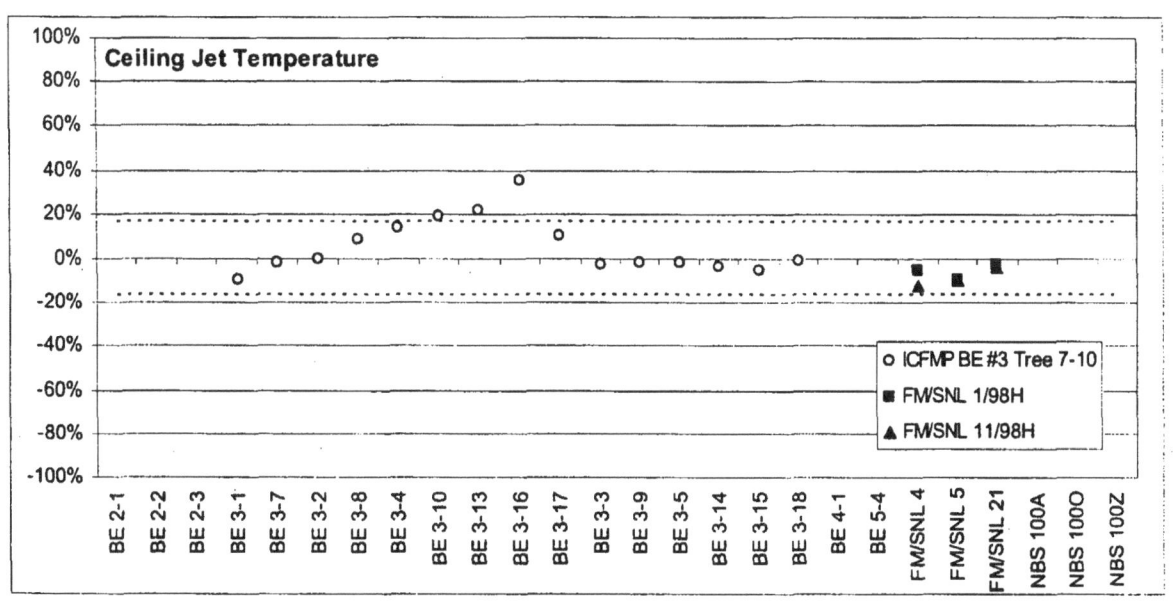

Figure 6-2: Scatter Plot of Relative Differences for Ceiling Jet Temperatures in ICFMP BE #3 and the Selected FM/SNL Tests (Experimental uncertainty is 16%.)

6.3 Plume Temperature

As with the ceiling jet, MAGIC has a specific plume sub-model. This validation refers primarily to the implementation of the McCaffrey plume temperature correlation and the correction for plume flows above the hot gas layer interface. The line plume model in MAGIC was not evaluated in this study. Data from ICFMP BE #2 and the FM/SNL test series have been used to assess the accuracy of plume temperature predictions.

ICFMP BE #2: MAGIC's predictions of plume temperature are within the experimental uncertainty of 14%. Figure A-22 provides the graphical comparisons between model predictions and experimental measurements. The calculated relative differences are listed in Table A-11 and plotted in Figure 6-3.

FM/SNL: MAGIC predicts the plume temperatures in Tests 4 and 5 to within experimental uncertainty. See Figure A-23 and Table A-12 for the graphical comparisons and calculated relative differences.

Summary: Plume Temperature — Green

The axisymmetric plume temperature model in MAGIC is well-suited for applications similar to those evaluated in this study.

- This V&V study included the evaluation of MAGIC capabilities for predicting plume temperature inside the hot gas layer.

- Because all of the relative differences are within experimental uncertainty and the experimental and predicted temperature profiles show good agreement, a classification of Green is assigned for the axisymmetric plume model in MAGIC.

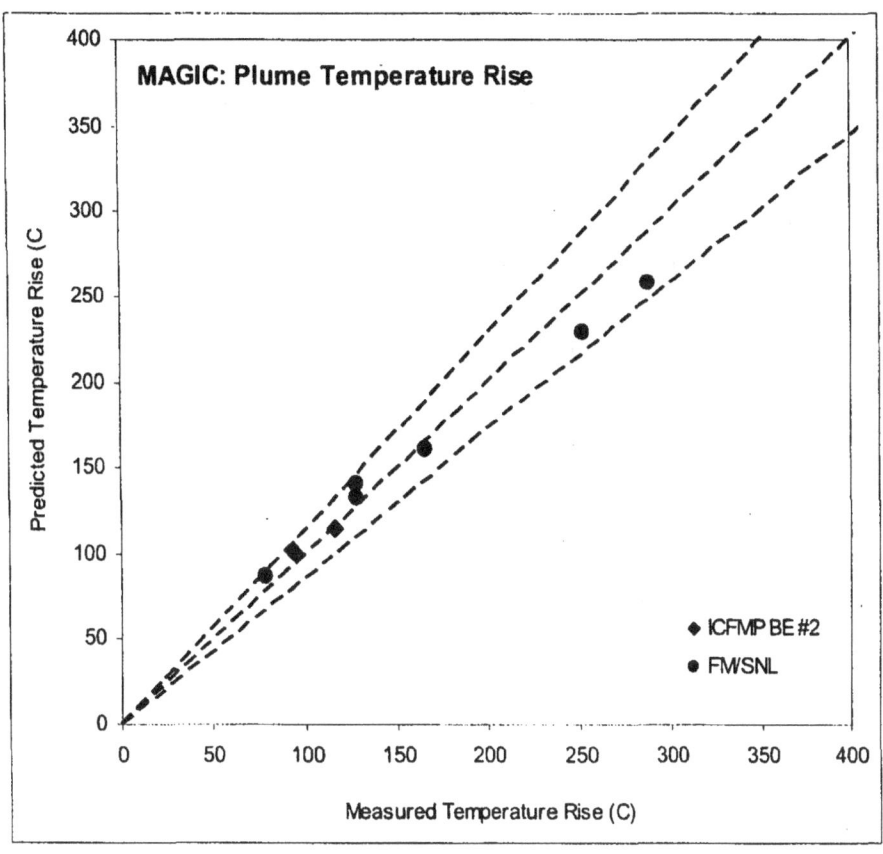

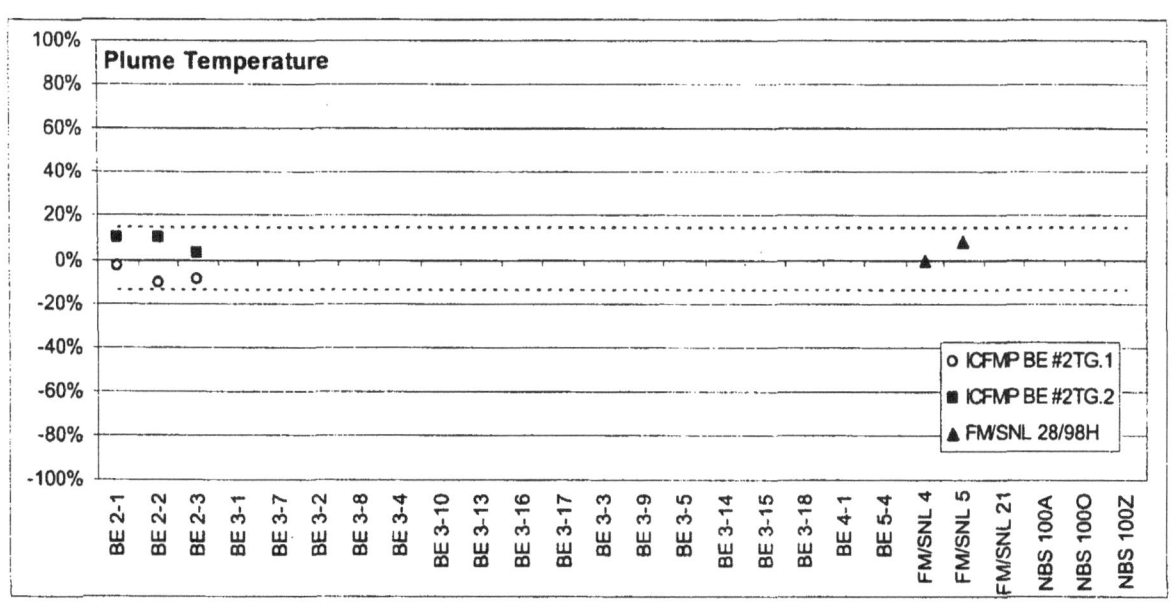

Figure 6-3: Scatter Plot of Relative Differences for Plume Temperatures in ICFMP BE #2, and the Selected FM/SNL Tests (Experimental uncertainty is 14%.)

6.4 Flame Height

Flame height is recorded by visual observations, photographs, or video footage. Videos from the ICFMP BE #3 test series and photographs from BE #2 are available. It is difficult to precisely measure the flame height, but the photos and videos allow one to make estimates accurate to within a pan diameter.

The MAGIC model for flame height consists of Heskestad's flame height correlation. See Reference 2 for technical details.

ICFMP BE #2: The height of the visible flame in the photographs of BE #2 has been estimated to be between 2.4 and 3 pan diameters [(3.8 m to 4.8 m (12.5 ft to 15.7 ft)]. From Figure A-24, which reports MAGIC flame height predictions, flame heights are between 3 and 7 m (9.8 and 23.0 ft).

ICFMP BE #3: MAGIC appears to predict the flame height correctly in this test series, at least to the accuracy of visual observations and a few photographs taken before the hot gas layer obscures the upper part of the fire. The experiments were not designed to measure the flame height other than through visual observation. Flame height pictures and MAGIC predictions can be found in Figures A-26 through A-28. Notice for example that Figure A-26 suggests flames with heights similar to the height of the door [2 m (6.6 ft)]. MAGIC's predictions peak above 2 m (6.6 ft) in all cases.

Summary: Flame Height — Green

MAGIC appears to provide flame height predictions consistent with the heights observed in available photographs for BE #2. MAGIC's flame height predictions for BE #3 are also consistent with the heights observed in available photographs.

- This evaluation does not suggest that MAGIC is under-predicting flame height. Therefore, based on the consistency with visual evidence, a Green classification is assigned.

6.5 Oxygen Concentration

The oxygen concentration in MAGIC results directly from the conservation of mass equation in both the upper and lower layers. The evaluation results are based on oxygen concentrations calculated in the upper layer. It should be stressed that this study is limited to well-ventilated fires (Equivalence ratios, ϕ, less than 1, as noted in Volume 1, Table 2-5).

ICFMP BE #3: The relative differences associated with MAGIC's predictions of upper-layer oxygen concentration range from approximately –30% to 25%. Some of these relative differences are outside the range of experimental uncertainty of 9%. As suggested in Figure 6-4, there appears to be a pattern of negative relative differences associated with open-door tests not observed in the closed-door tests. In all these cases, the measured oxygen concentration was above 15%. In terms of the closed-door tests, all the relative differences are within experimental uncertainty with the exception of Tests 4 and 10, which involved a mechanically ventilated room. Recall that negative relative differences indicate that MAGIC predicted oxygen concentrations higher than those measured in the experiments. Figures A-29 and A-30 illustrate the experimental and model oxygen concentration profiles.

ICFMP BE #5: MAGIC's prediction of the upper-layer oxygen concentration in Test 4 of this test series is above the experimental uncertainty of 9%.

Summary: Oxygen Concentration — Yellow

The MAGIC model is capable of making oxygen concentration predictions, assuming that the basic stoichiometry of the combustion reaction is known. Recall that this study is limited to well-ventilated compartment fires only.

- Relative differences in BE #3 are comparable to experimental uncertainty in the case of closed-room tests. In the case of open-door tests, they are below the experimental uncertainty range in BE #3. The relative difference associated with the single comparison in BE #5 is also outside the range of experimental uncertainty.

- Based on the above discussion, a classification of Yellow is assigned for the oxygen concentration predictions in MAGIC.

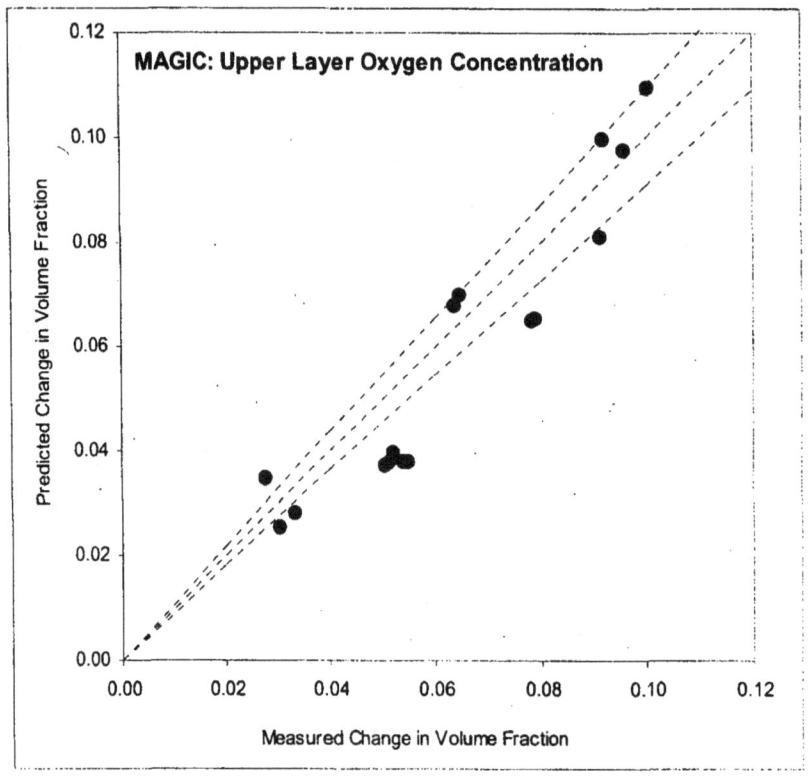

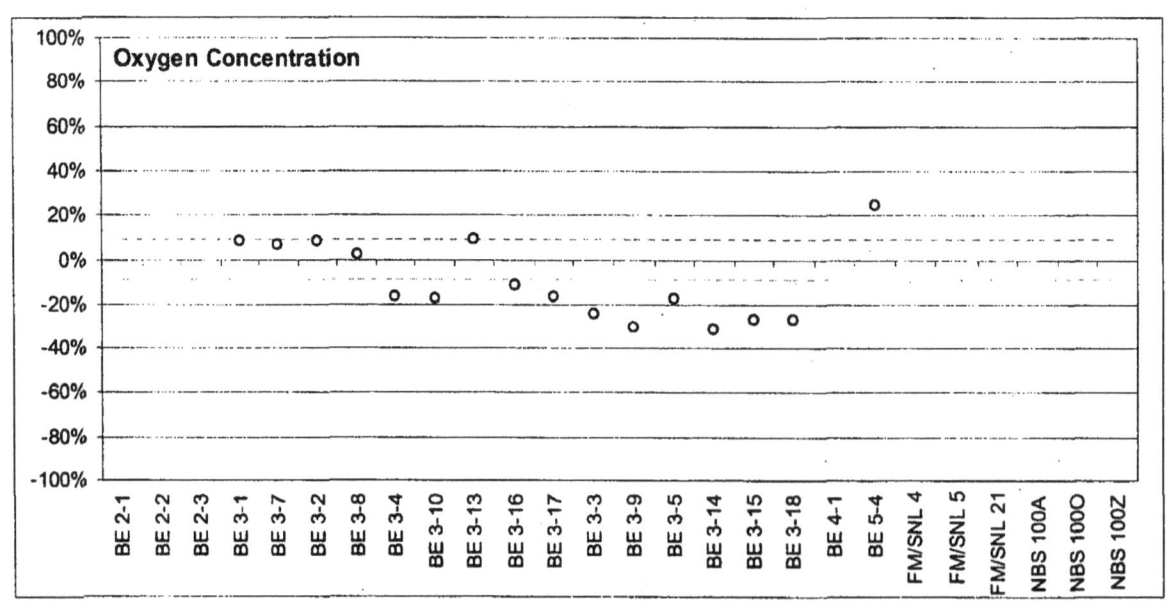

Figure 6-4: **Scatter Plot of Relative Differences for Oxygen Concentration in ICFMP BE #3 and #5**
(Experimental uncertainty is 9%.)

6.6 Smoke Concentration

Only ICFMP BE #3 has been used to assess predictions of smoke concentration. For these tests, the smoke yield was specified as one of the test parameters. MAGIC consistently over-predicted smoke concentrations from approximately 30% to 500% in the closed-door tests. For open-door tests, MAGIC's predictions are within the experimental uncertainty of 33%.

The graphical comparisons for smoke concentration are summarized in Figures A-32 and A-33. The relative differences are listed in Table A-15 and plotted in Figure 6-5.

Summary: Smoke Concentration — Yellow

MAGIC is capable of transporting smoke throughout a compartment, assuming that the production rate is known and its transport properties are comparable to gaseous exhaust products.

- MAGIC over-predicts the smoke concentration in closed-door tests. The predictions for open-door tests are within experimental uncertainty.

- No firm conclusions can be drawn to explain the drastic differences in predictions between open- and closed-door tests. Therefore, a Yellow classification is assigned because there is no clear indication that MAGIC would always result in conservative estimates.

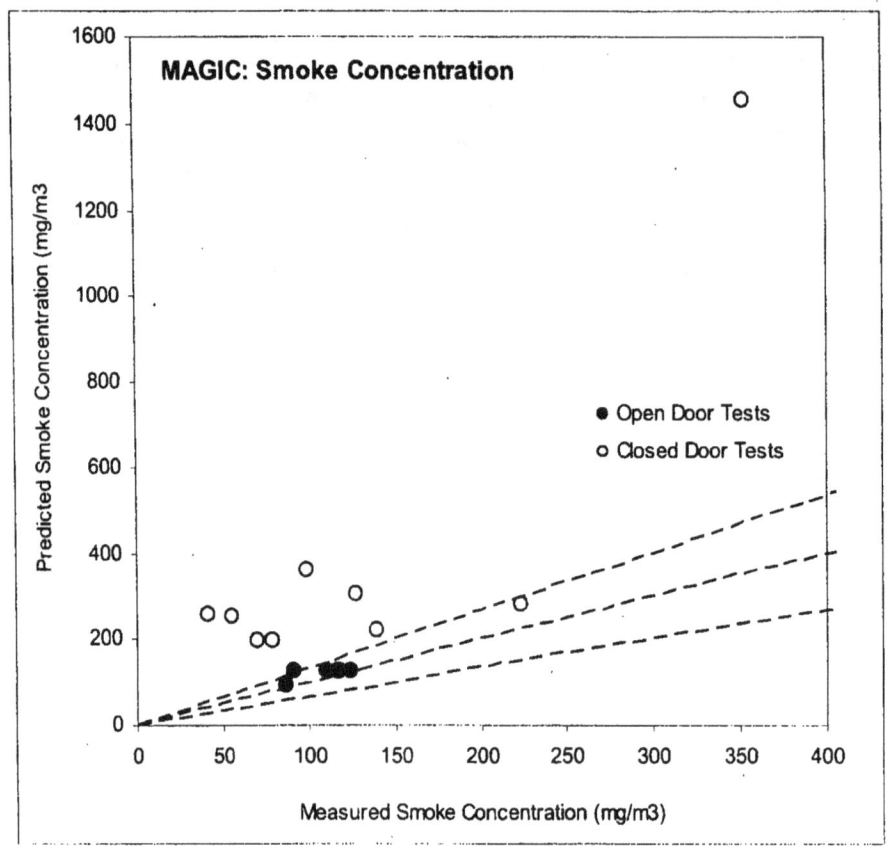

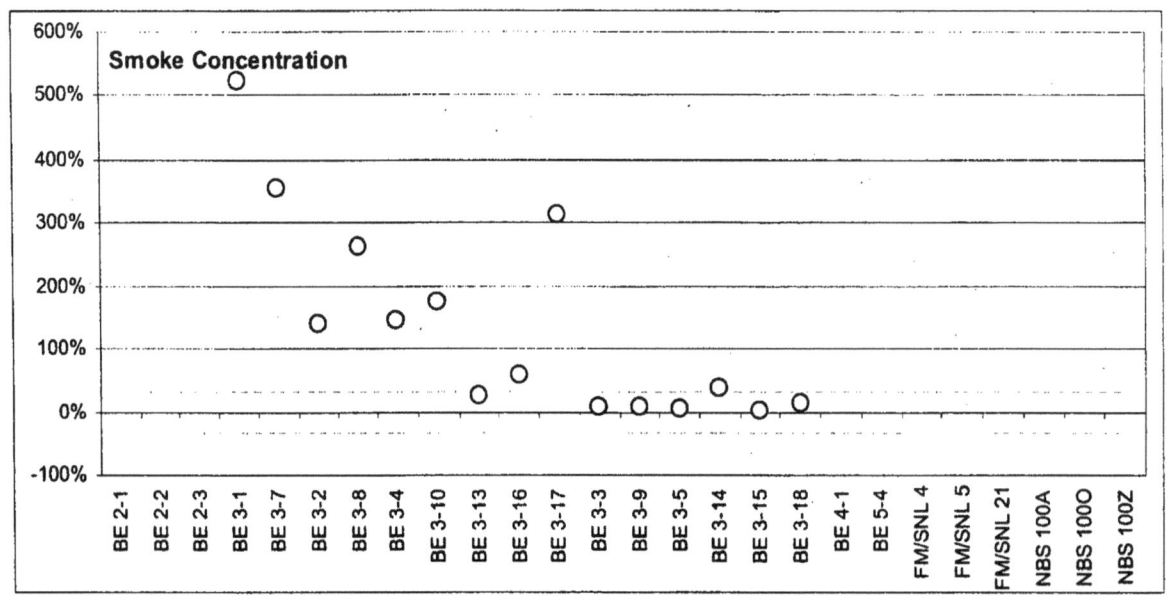

Figure 6-5: Scatter Plot of Relative Differences for Smoke Concentration in ICFMP BE #3 (Experimental uncertainty is 33%.)

6.7 Compartment Pressure

Comparisons between measurements and predictions of compartment pressure for BE #3 are shown in Figure A-34 and A-35. For those tests in which the door to the compartment is open, the over-pressures are only a few Pascals; however, when the door is closed, the over-pressures are several hundred Pascals.

The relative differences were calculated as follows:

- For closed-door rooms, the relative difference refers to the positive peak at the early stages of the fire. Positive relative differences indicate that MAGIC over-predicted the measured peak.

- For open-door rooms, the relative difference refers to the negative magnitudes of the pressure, typically at the late stages of the test. Positive relative differences suggest that MAGIC calculated the difference as being lower than the experimental measurement.

Relative differences are listed in Table A-16.

Visual examination of the plots of experimental data and model results (see Figure A-34) strongly suggest that for tests with open doors, where leakages are not critical because of the large door opening, MAGIC captures both the magnitude and the profile of the pressure. These profiles suggest a negative pressure profile at the floor of the room, indicating that fresh air is moving into the enclosure.

In closed-door tests, MAGIC captures both peaks and pressure profiles (see Figure A-35). It is important to mention that fan tests were conducted before some of the tests resulting in relatively well-known leakage areas, which were used as inputs to the model. Furthermore, notice that MAGIC captures both positive and negative pressure peaks. These peaks are an indication of a positively pressurized room in the early stages of the test, and a negatively pressurized room when the fuel supply is discontinued and heat losses to the boundaries are higher than the heat release rate of the fire.

In general, the magnitudes of the predicted pressures are comparable to those of the measured pressures and, in most cases, differences can be explained using the reported uncertainties in the leakage area and the fact that the leakage area changed from test to test because of the thermal stress on the compartment walls. The one notable exception is Test 16, which was performed with the door closed and the ventilation on. For that test, there is considerable uncertainty in the magnitudes of both the supply and exhaust flow rates.

The relative differences are plotted in Figure 6-6. Notice that only the relative difference associated with Test 16 is outside the experimental uncertainty ranges of 40% and 80% for tests with the ventilation system off or on, respectively.

Summary: Compartment Pressure — Green

- The basic mass and energy conservation equations solved by MAGIC ensure reliable predictions of compartment pressure. It should be stressed that compartment pressure predictions are extremely sensitive to the leakage area and forced ventilation. In the MAGIC runs, the leakage area listed in the experimental descriptions was divided by the orifice flow coefficient, 0.68, so that it is reflected in the model as the actual opening area.

- The MAGIC pressure predictions for BE #3 are within experimental uncertainty, with an exception that may be related to the behavior of a ventilation fan.

- A Green classification is assigned for compartment pressure predictions in MAGIC, assuming that room leakages are known in closed-door scenarios.

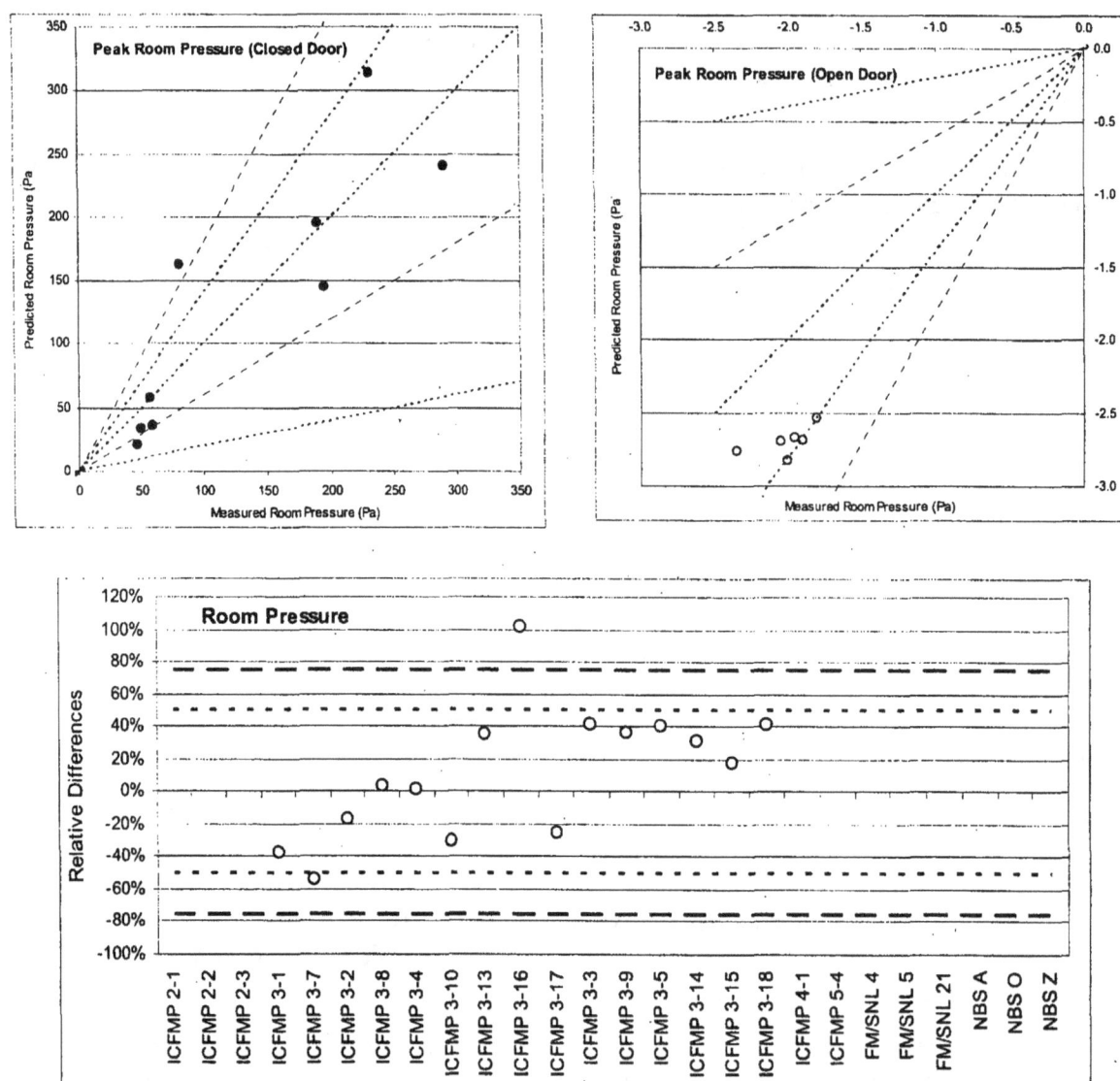

Figure 6-6: Scatter Plot of Relative Differences for Room Pressure in ICFMP BE #3 [Experimental uncertainties are 40% (no forced vent) and 80% (forced vent).]

6.8 Target Temperature and Heat Flux

Target temperature and heat flux data are available from ICFMP BE #3, #4, and #5. In BE #3, the targets are various types of cables in various configurations — horizontal, vertical, in trays, or free-hanging. In BE #4, the targets are three rectangular slabs of different materials instrumented with heat flux gauges and thermocouples. In BE #5, the targets are again cables, in this case, bundled power and control cables in a vertical ladder.

ICFMP BE #3: There are nearly 200 comparisons of heat flux and surface temperature on four different cables that are graphed in Section A.8.1. Consequently, it is difficult to make sweeping generalizations about the accuracy of MAGIC. For each target, the graphical comparison of experimental measurements and MAGIC predictions are presented for target temperature, cable temperature, radiation, and total heat flux. At best, one can scan the figures and associated tables to get a sense of the overall performance. The experimental uncertainty is about 20% and 14% for heat flux and surface temperature, respectively. The following important aspects of this evaluation should be considered:

- MAGIC provides the capability to model cable temperature as a "target" (where the material is simulated as a slab), or a cable itself (where the material is simulated as a cylinder with concentric layers of conductor, insulation, and jacket). This evaluation includes both alternatives. When cables were modeled as targets (slab), the thickness of the target was selected as d/4, where d is the diameter of the cable (see the discussion on cable modeling and sensitivity analysis in Chapters 3 and 5).

- The measured radiative heat flux is compared with MAGIC's "Incident Heat Flux" output, which is the sum of all radiated heat fluxes to a target. That is, the incident flux is the radiation flux received from the environment. This value is normally positive because it is not a balance. At the beginning of the simulation, conditions are ambient. Therefore, the incident heat flux is approximately $\varepsilon \sigma T_{amb}^4 \approx 0.9(5.67E-11)(293)^4 \approx 0.4 \, kW/m^2$.
  The total heat flux measurement is compared with MAGIC's "Total Heat Flux, Flux Meter" output, which simulates a typical water-cooled heat flux meter.

Figures 6-7 through 6-10 show the relative differences for target and cable temperature, as well as radiative and total heat fluxes for targets B-TS-14, D-TS-12, F-TS-20, and G-TS-33. The following observations are relevant:

- It can be concluded that results from the majority of the comparisons were within experimental uncertainty or were over-predictions for surface temperature.

- In general, there is more scatter in heat flux predictions than in surface temperature predictions.

- In the case of temperature, there is almost no difference in modeling cables as targets or cables, provided that the thermo-physical properties are the same and the thickness of the target is one-quarter of the diameter of the cable.

- Relative differences for heat flux suggest under- and over-predictions.

- Specific conclusions can be drawn on a case-by-case basis. For example, temperatures for G-TS-33 are generally over-predicted.

ICFMP BE #4: MAGIC over-predicts both the heat flux and surface temperature of three "slab" targets located about 1 m (3.3 ft) from the fire. The trend is consistent, but it cannot be explained solely in terms of experimental uncertainty. The technical details supporting relative differences are included in Figure A-68 and Table A-21.

ICFMP BE #5: MAGIC predicts both temperature and total heat flux to targets in BE #5, Test 4 (approximately) to within experimental uncertainty. The technical details supporting relative differences are included in Figure A-69 and Table A-22.

<u>**Summary:**</u> **Radiant and Total Heat Flux and Target Temperature — Yellow**

MAGIC is capable of predicting the radiative and total heat flux to targets, assuming known thermo-physical properties. MAGIC is also capable of predicting the surface temperature of a target. Based on the scatter plots of relative differences, the following classifications are assigned:

- Yellow for surface temperature of the target
- Yellow for total heat flux
- Yellow for radiated heat flux

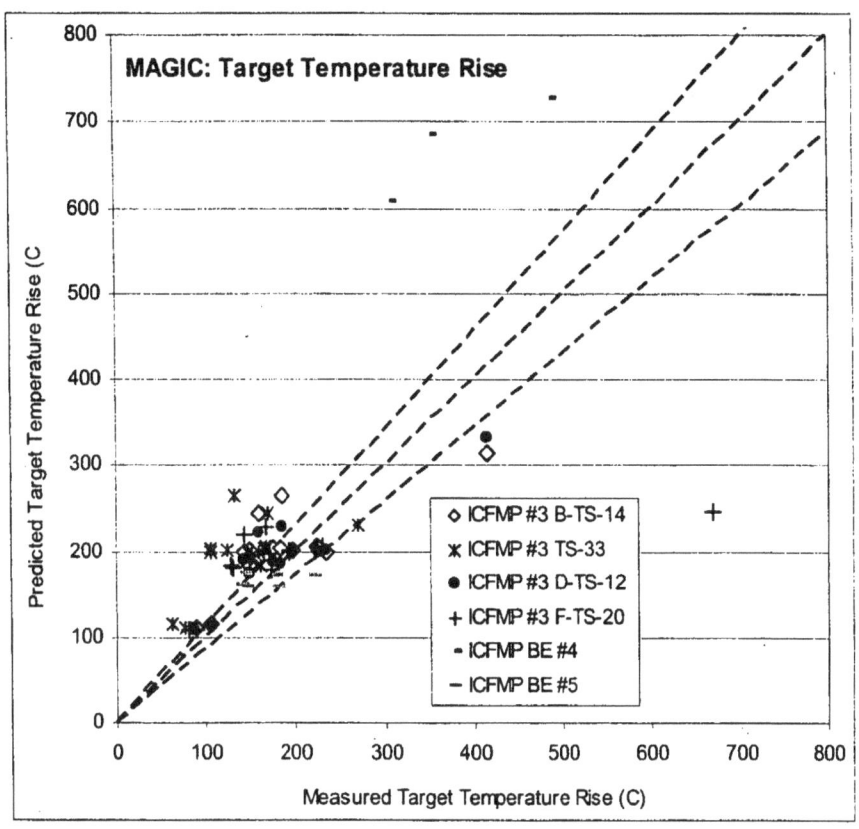

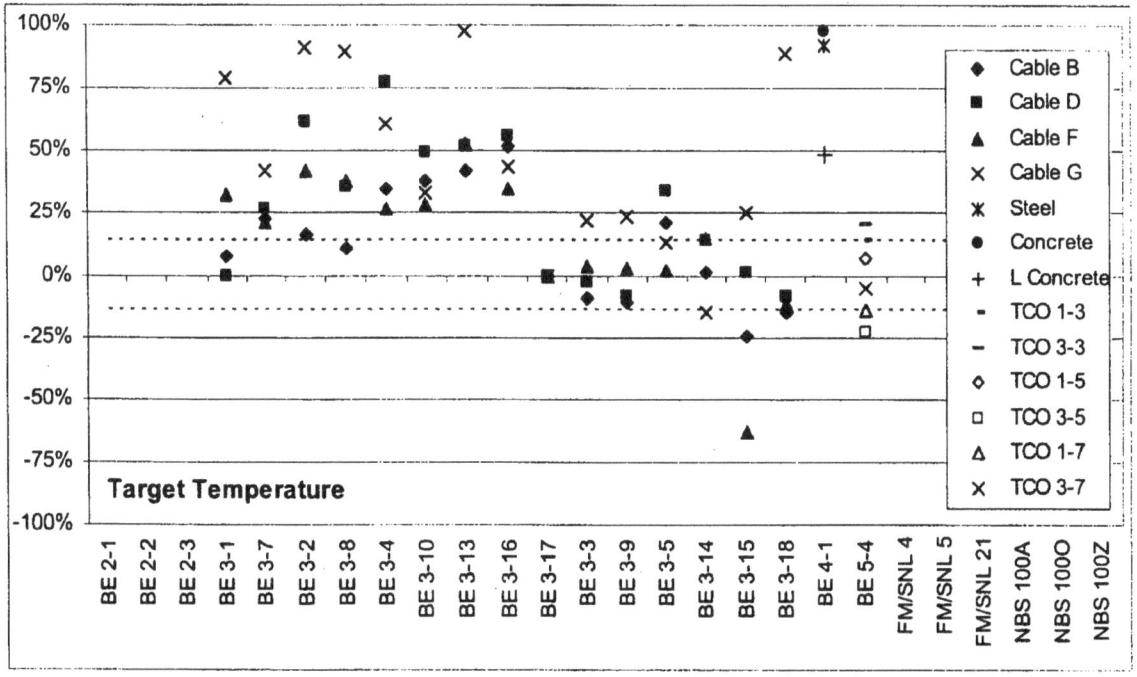

**Figure 6-7: Scatter Plot of Relative Differences for Target Temperature in ICFMP BE #3
(Experimental uncertainty is 14%.)**

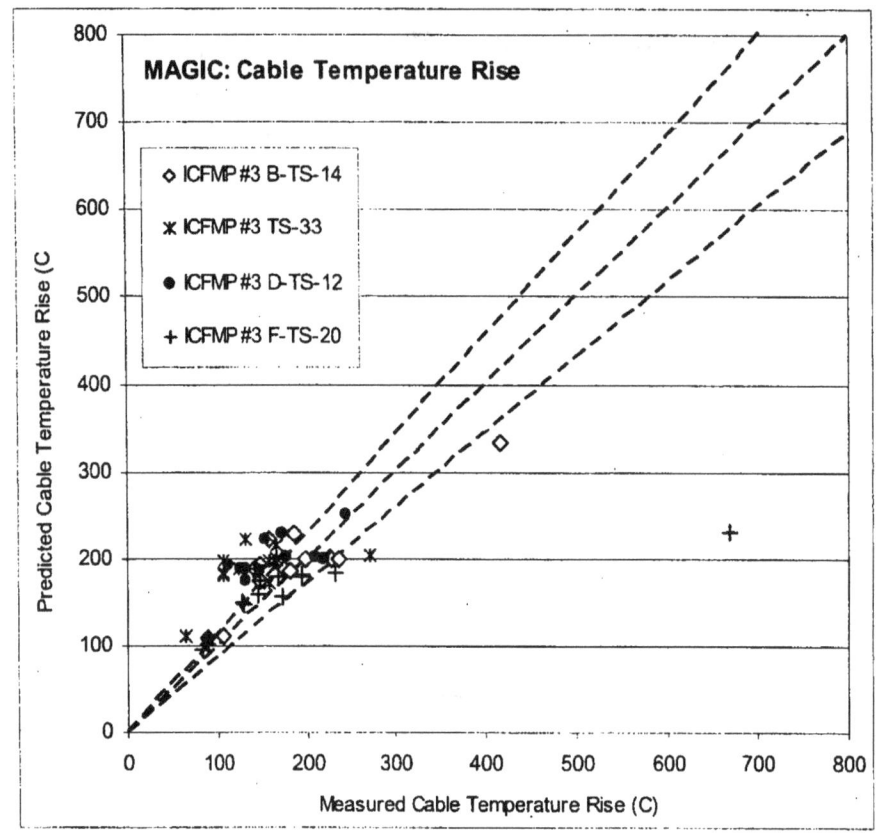

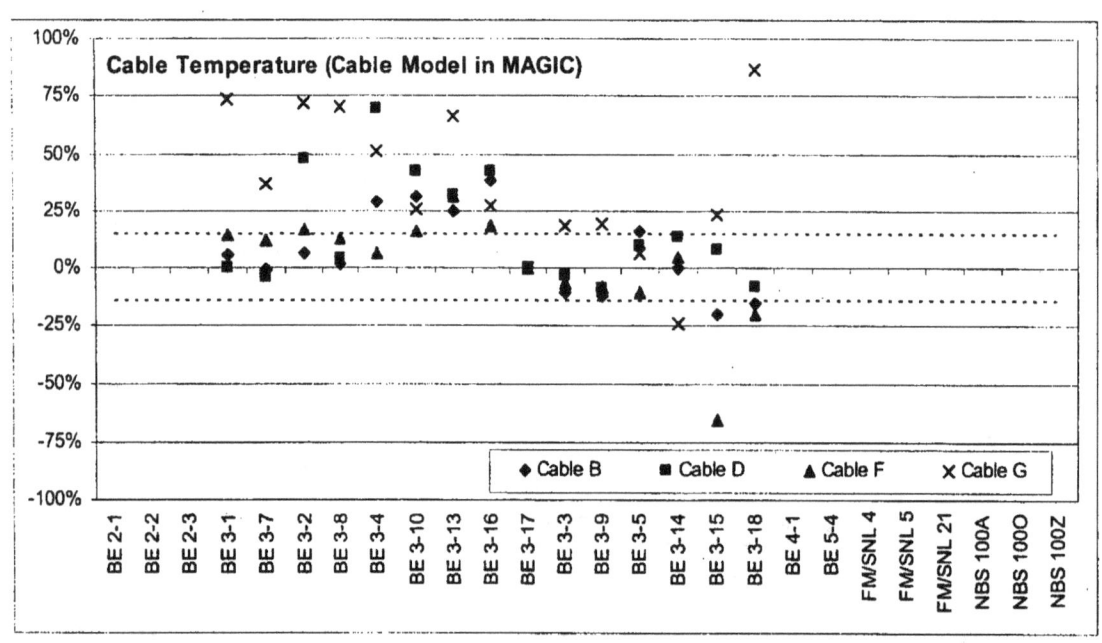

Figure 6-8: Scatter Plot of Relative Differences for Target Temperature in ICFMP BE #3 (Experimental uncertainty is 14%.)

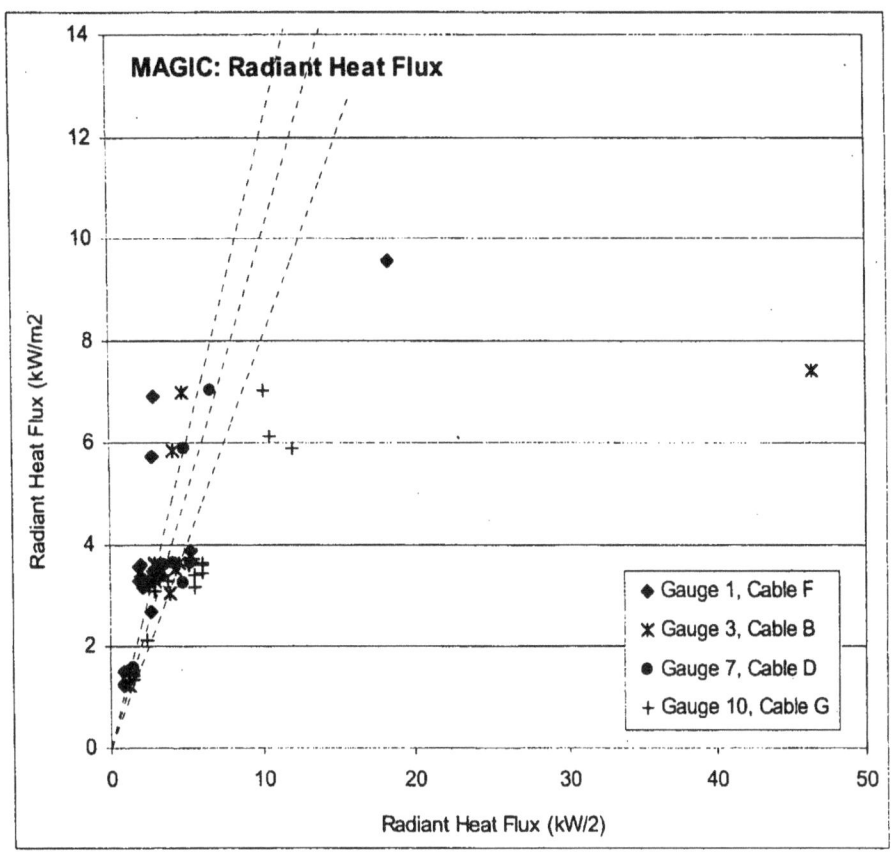

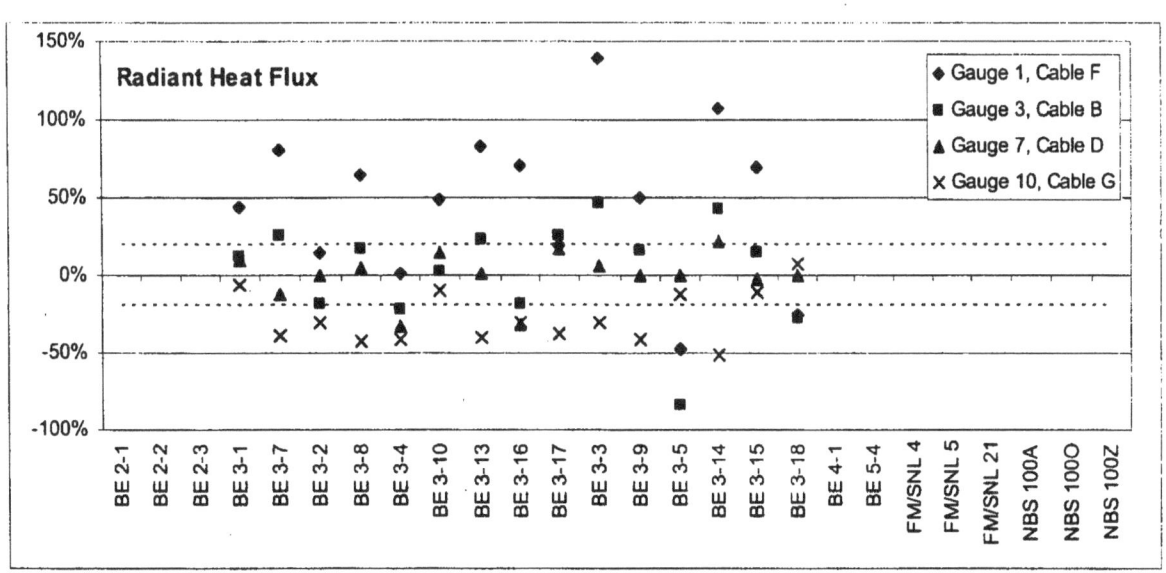

Figure 6-9: Scatter Plot of Relative Differences for Radiant Heat Flux in ICFMP BE #3 (Experimental uncertainty is 20%.)

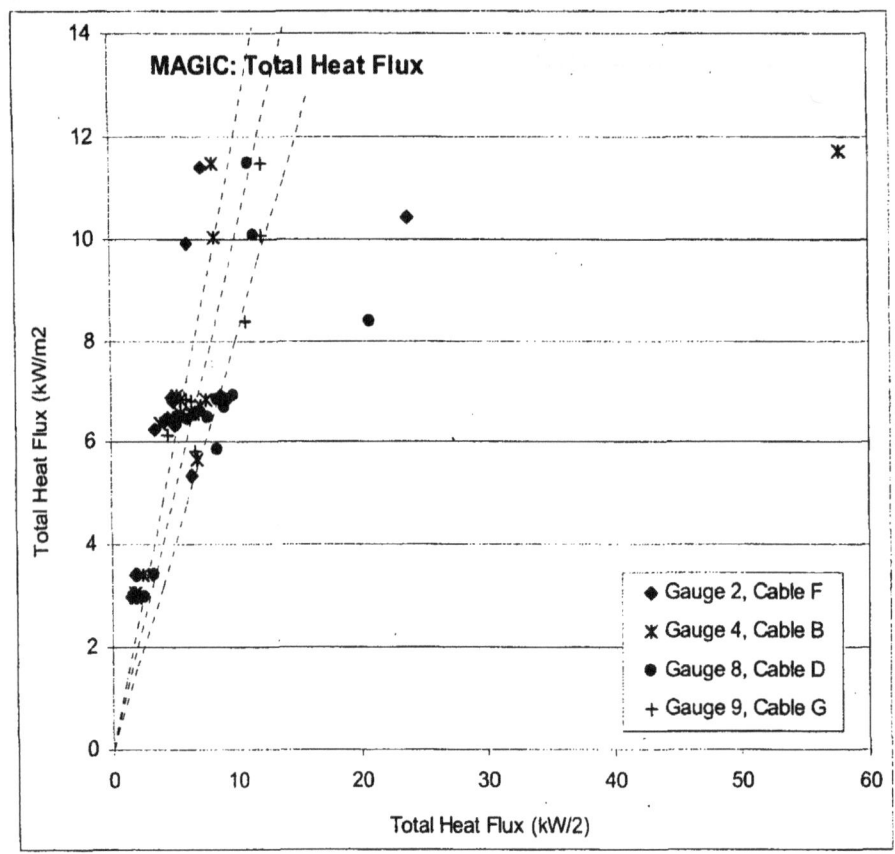

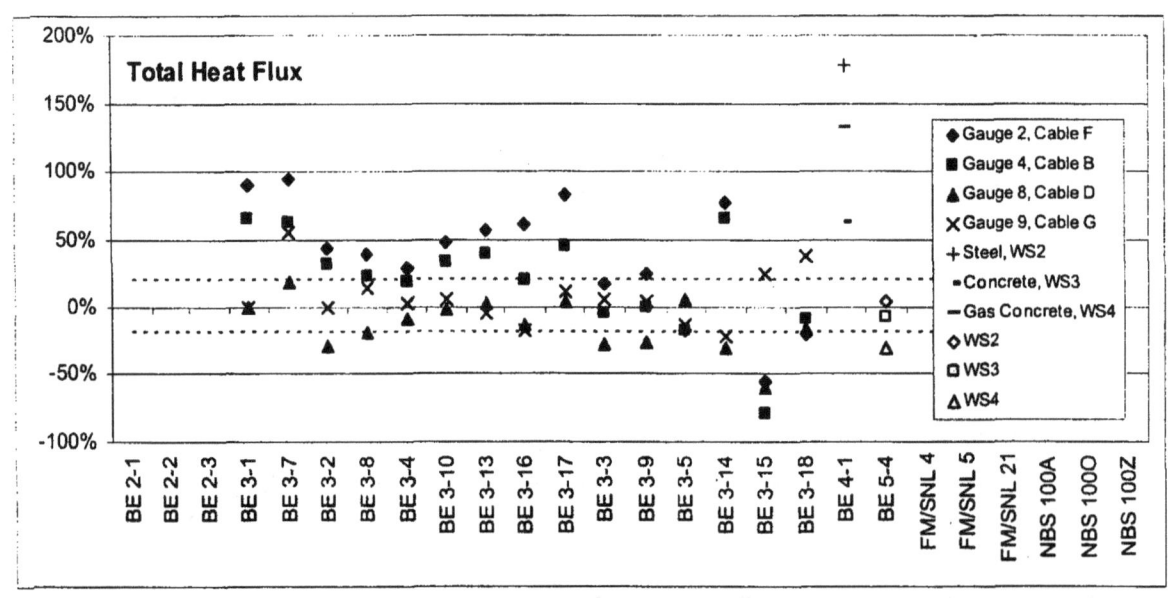

Figure 6-10: Scatter Plot of Relative Differences for Total Heat Flux in ICFMP BE #3 (Experimental uncertainty is 20%.)

6.9 Wall Heat Flux and Temperature

Wall heat flux and surface temperature measurements are available from ICFMP BE #3, and wall surface temperature measurements are also available from BE #4 and BE #5. As with target heat flux and surface temperature (above), there are numerous comparisons.

It should be noted that the wall temperatures and heat fluxes in MAGIC were calculated locating individual targets in the walls. The targets are characterized by the thermo-physical properties and thickness of the wall. The targets were located in the same model location as the experimental instruments in the test room. Consequently, this evaluation does not include MAGIC's "Wall Temperature," or "Wall Heat Flux" output options, which are available in the Wall output category. Experimental measurements were compared with MAGIC's "Total Absorbed Heat Flux" output option.

ICFMP BE #3: It cannot be generalized that MAGIC predicts wall temperatures and heat fluxes within the experimental uncertainty of 14% and 20%, respectively. For the most part, temperatures and heat fluxes for walls are over-predicted, while those for ceilings and floors present both under and over predictions. As noted by the corresponding markers in Figures 6-11 and 6-12, most of the relative differences for the ceiling and floor temperatures and heat fluxes are negative. The over-predictions for walls and floors can be up to (approximately) 100% with very few exceptions.

The graphical comparisons of experimental and predicted temperature and heat flux profiles are presented in Figures A-70 through A-85 and Tables A-24 through A-27.

ICFMP BE #4: MAGIC predicted two wall surface temperatures to within the experimental uncertainty of 20%. The two points are presumably very close to the fire because the temperatures are 600 °C to 700 °C (see Figure A-86) above ambient. The relative differences are –11% and 10%, as listed in Table A-28.

ICFMP BE #5: MAGIC predictions of wall temperature are comparable to experimental uncertainty with a significant outlier of more than 800%. At this point, there is no explanation for such an outlier.

<u>Summary</u>: **Wall Heat Flux and Temperature — Yellow**

MAGIC has the capability to predict the radiative and total heat flux to walls. MAGIC also has the capability to predict the surface temperature of a wall, assuming that its composition is fairly uniform and its thermal properties are well-characterized.

- MAGIC generally over-predicted the heat flux and surface temperature for walls and floors, with few comparisons below the lower limit of experimental uncertainty. By contrast, MAGIC consistently under-predicted the heat flux and surface temperature for ceilings. Based on these results, a Yellow classification is assigned.

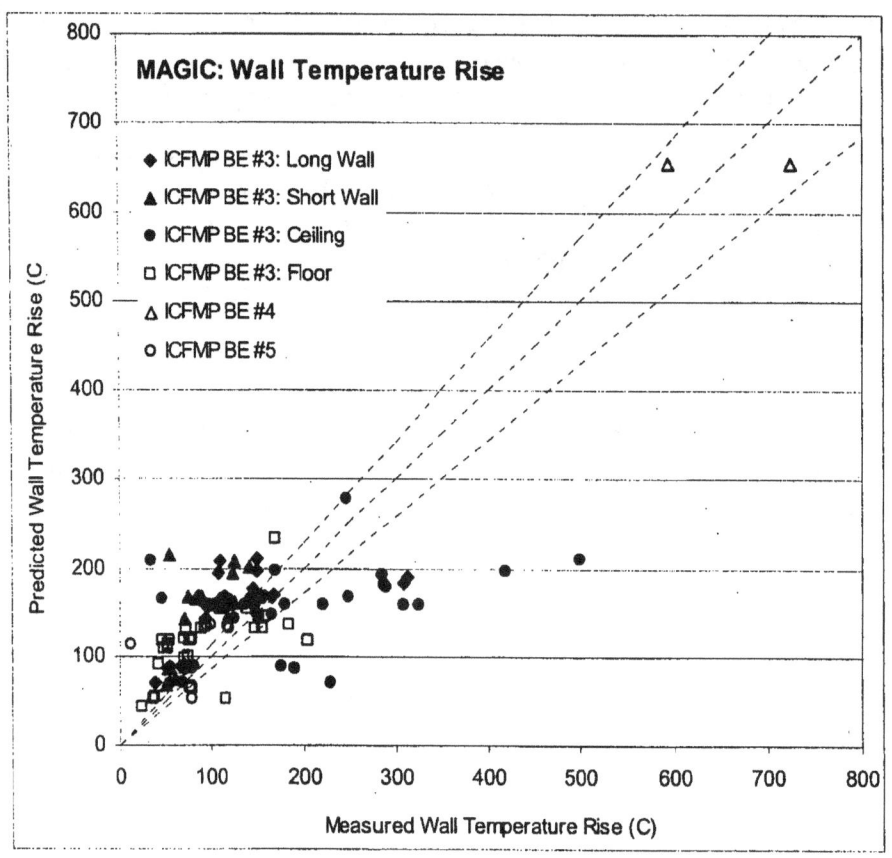

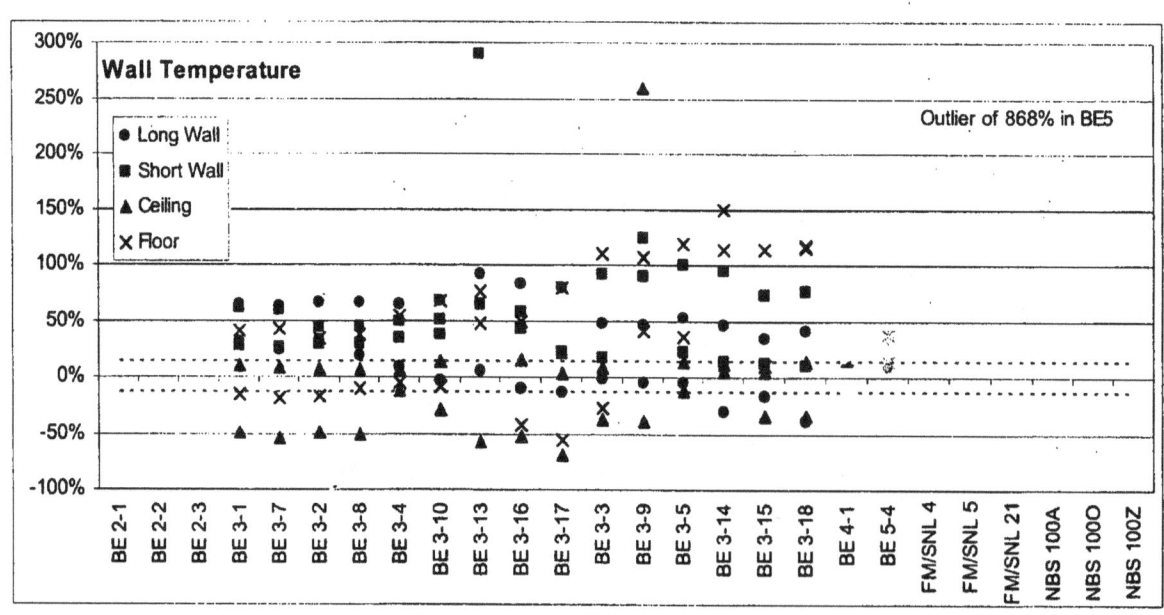

Figure 6-11: Scatter Plot of Relative Differences for Target Temperature in ICFMP BE #3, #4, and #5 (Experimental uncertainty is 14%.)

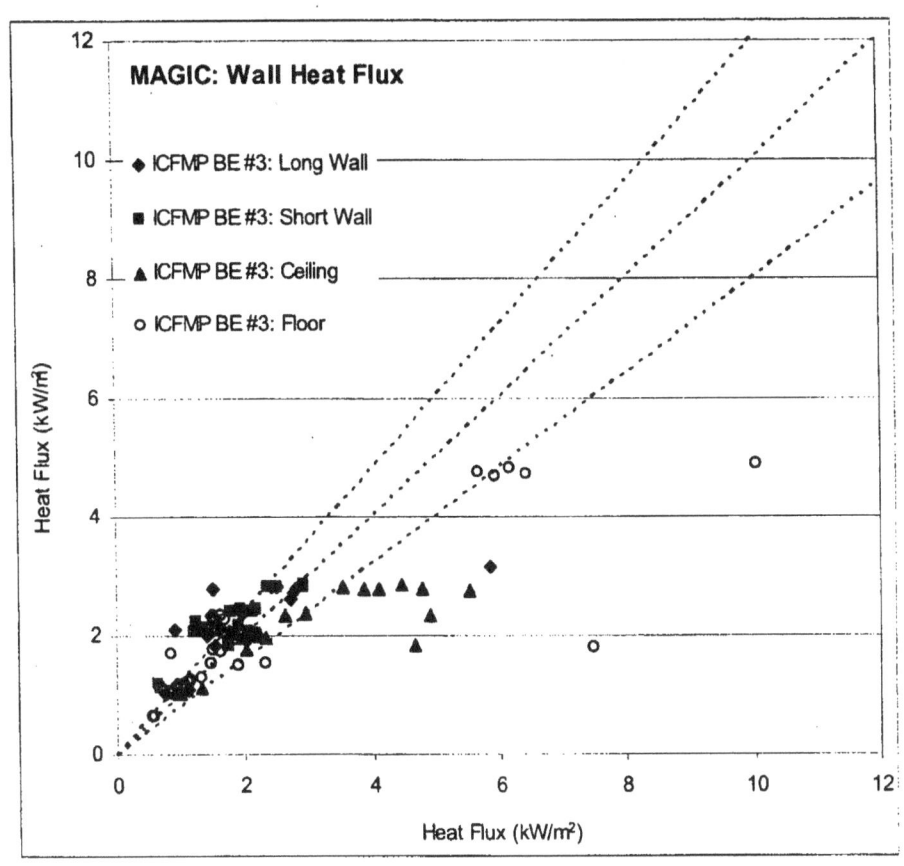

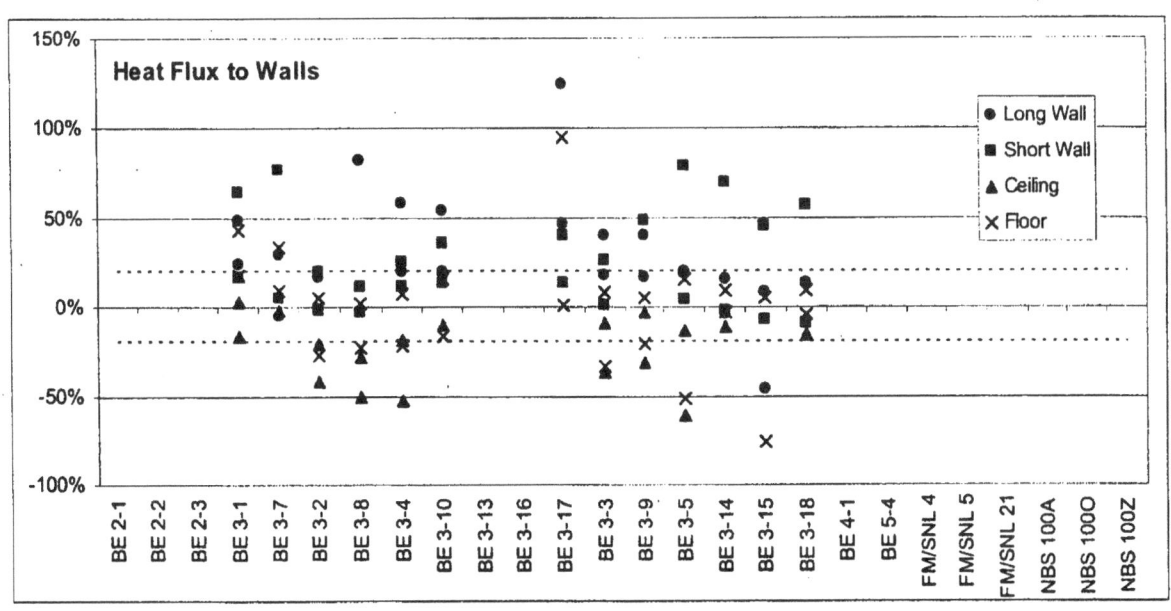

Figure 6-12: Scatter Plot of Relative Differences for Heat Flux in ICFMP BE#3, #4 and #5 (Experimental uncertainty is 20%.)

6.10 Summary

This chapter summarizes numerous comparisons of the MAGIC output with a range of experimental results obtained as part of this V&V effort. Thirteen quantities were selected for comparison and a color rating was assigned to each output category, indicating, in a very broad sense, how well the model treats that particular quantity:

- HGL Temperature and Height: Green
- Ceiling Jet Temperature: Green
- Plume Temperature: Green
- Flame Height: Green
- Oxygen: Yellow
- Smoke Concentration: Yellow
- Compartment Pressure: Green
- Radiation Heat Flux, Total Heat Flux, and Target Temperature: Yellow
- Wall Heat Flux and Surface Temperature: Yellow

Five of the quantities were assigned a Green rating, indicating that the research team concluded that the physics of the model accurately represent the experimental conditions, and the calculated relative differences (comparing the model output and experimental results) are consistent with the combined experimental and input uncertainty. A few notes on the comparisons are appropriate:

- The MAGIC predictions of HGL temperature and height are, with a few exceptions, within or close to experimental uncertainty.

- MAGIC's predictions for ceiling jet and plume temperatures are comparable to experimental uncertainty. In the case of the ceiling jet, results suggest a greater scatter for relative difference in BE #3 closed-door tests than for BE #3 open-door tests. At this point, no specific explanation for this behavior is available. In the case of plume temperature, all relative differences were within experimental uncertainty.

- MAGIC predicts the flame height consistent with visual observations of flame height for the experiments. This is not surprising, given that MAGIC simply uses a well-characterized experimental correlation to calculate flame height.

- Compartment pressure: MAGIC predicted compartment pressure to within experimental uncertainty.

Four of the quantities were assigned a Yellow rating, indicating that users should exercise caution when using the model to evaluate the given quantity. This typically indicates limitations in the use of the model. A few notes on the comparisons are appropriate:

- MAGIC generally over-predicts smoke concentration. Predicted concentrations for open-door tests are within experimental uncertainties, but those for closed-door tests are far higher.

- MAGIC predicts most cable surface temperatures within or above experimental uncertainties. Very few under-predictions were observed. However, this is not the case for total and radiant heat fluxes. Relative differences are both under- and over-predicted. Total heat flux to targets is typically predicted to within about 30%, and often under-predicted. Care should be taken in predicting localized conditions (such as target temperature and heat flux) because of inherent limitations in all zone fire models.

- Oxygen concentrations were consistently under-predicted at about 30% for open door tests in BE #3. However, these under-predictions resulted from oxygen concentration comparisons above 15%, which are above concentrations suggesting fire extinction. In the case of closed-door tests, MAGIC's results are comparable to experimental uncertainty.

- MAGIC generally over-predicted the compartment surface temperature and heat flux for walls, but consistently under-predicted the surface temperature and heat flux for ceilings. Finally, the floor surface presents both over and under-predictions.

- Differences between the model outputs and experimental results were evident in these studies. Some of the differences can be explained by limitations of the model and/or the experiments. Like all predictive models, the best predictions come with a clear understanding of the limitations of the model and the inputs provided for the calculations.

7
REFERENCES

1. *ASTM Standard Guide for Evaluating the Predictive Capability of Deterministic Fire Models*, ASTM E 1355-05a, American Society for Testing and Materials, West Conshohocken, PA, 2005.

2. Gay, L., C. Epiard, and B. Gautier, *MAGIC Software Version 4.1.1: Mathematical Model*, EdF HI82/04/024/B, Electricité de France, France, November 2005.

3. Gay, L., *User Guide of the MAGIC Software V4.1.1"* EdF HI82/04/23/A, Electricité de France, France, April 2005.

4. Gay, L., J. Frezabeu, and B. Gautier, *Qualification File of Fire Code MAGIC Version 4.1.1*, EdF HI-82/04/022/B, Electricité de France, France, November 2005.

5. McCaffrey, B., "Flame Height," *The SFPE Handbook of Fire Protection Engineering*, 2nd Ed. (P.J. DiNenno, D. Drysdale, C.L. Beyler, and W.D. Walton, eds.), National Fire Protection Association and The Society of Fire Protection Engineers, Quincy, MA, 1995.

6. Heskestad, G., "Fire Plume, Flame Height, and Air Entrainment," *The SFPE Handbook of Fire Protection Engineering*, 3rd Ed. (P.J. DiNenno, D. Drysdale, C.L. Beyler, and W.D. Walton, eds.), National Fire Protection Association and The Society of Fire Protection Engineers, Quincy, MA, 2002.

7. Cooper, L., "Fire Plume-Generated Ceiling Jet Characteristics and Convective Heat Transfer to Ceiling and Wall Surfaces in a Two-Layer Fire Environment: Uniform Temperature, Ceiling and Walls," *Fire Science & Technology*, Vol 13, No. 1 & No. 2, pp. 1–17, 1993. Also NISTIR 4705, November 1991.

8. Cooper, L.Y., "Combined Buoyancy and Pressure-Driven Flow through a Horizontal Vent," NISTIR 5384, National Institute of Standards and Technology, Gaithersburg, MD,

9. Heskestad, G., "Quantification of Thermal Responsiveness of Automatic Sprinklers Including Conduction Effects," *Fire Safety Journal*, Vol. 14, 1988.

10. Gear, R., *Dif sub for solution of Ordinary Differential Equations*, Collected algorithms from CACM, Algorithm 407, 1970.

11. Press, W.H., S.A. Teukolsky, W.H. Vetterling, and B.P. Flannery, *Numerical Recipes in FORTRAN*, Cambridge University Press, Cambridge, England, 1992.

12. Benmamoun, A., *"Rapport d'analyse et des modifications du code FORTRAN"* SYSAM - SE-0310AB, 2004 (*Analysis and modification report of the FORTRAN code*, in French).

13. FOR_STUDY (FORTRAN Code Analyzer), Cobalt Blue, Inc., Alpharetta, Georgia, available at http://www.cobalt-blue.com/fy/fymain.htm as of the time of publication of this report.

14. plusFORT Version 6 (Toolkit for FORTRAN Programmers), Polyhedron Software, Ltd., Standlake, England, available at http://www.polyhedron.com/pf/plusfort.html as of the time of publication of this report.

15. Gautier, B., *Plan Qualité du Logiciel MAGIC,* Note EdF/DER HT 31/95/025/B, Electricité de France, France, November 1996 (Quality assurance policy, MAGIC).

16. Casal, J., and T. Noguer, *MAGIC Version 4 Dossier de conception globale - Mise à jour,* Doc. ILM-technologie, September 2004 (Concept document, MAGIC user interface, in French).

17. Gay, L., and J. Frezabeu, *Cahier de recette de l'interface utilisateur du logiciel MAGIC en version 4.1.1,* EdF HI-82/04/021/P, Electricité de France, France, November 2004 (Magic user interface, in French).

18. Gay, L., J. Casal, and T. Noguer, *Manuel de référence de l'interface utilisateur du logiciel MAGIC en version 4,* HI-82/04/025/P, September 2004 (Reference handbook, MAGIC user interface, in French).

19. NUREG-1758, "Evaluation of Fire Models for Nuclear Power Plant Applications: Cable Tray Fires," U.S. Nuclear Regulatory Commission, Washington, DC, June 2002.

20. Barakatm M., "Interaction Rayonnment-Particules.Cas de Fummes Generees par Differents Types de Combustibles," Thesis from the University of Poitiers, Poitiers, France, 1994 (*Particle-Radiation Interaction, Case of Smokes Generated by Different Combustible Types,* in French).

21. Mulholland, G., Section 2–13, "Smoke Production and Properties" *The SFPE Handbook of Fire Protection Engineering,* 3rd Ed. (P.J. DiNenno, D. Drysdale, C.L. Beyler, and W.D. Walton, eds.), National Fire Protection Association and The Society of Fire Protection Engineers, Quincy, MA, 2002.

A
TECHNICAL DETAILS FOR THE MAGIC VALIDATION STUDY

This appendix provides comparisons of MAGIC's predictions and experimental measurements for the six series of fire experiments under consideration. The sections to follow contain assessments of the model's predictions for the following quantities:

A.1   Hot Gas Layer Temperature and Height
A.2   Ceiling Jet Temperature
A.3   Plume Temperature
A.4   Flame Height
A.5   Oxygen Concentration
A.6   Smoke Concentration
A.7   Compartment Pressure
A.8   Target Temperature and Heat Flux
A.9   Wall Heat Flux and Temperature

The model's predictions are compared to the experimental measurements in terms of the relative difference between the maximum (or where appropriate, minimum) values of each time history:

$$\varepsilon = \frac{\Delta M - \Delta E}{\Delta E} = \frac{\left(M_p - M_o\right) - \left(E_p - E_o\right)}{\left(E_p - E_o\right)}$$

ΔM is the difference between the peak value of the model prediction, M_p, and its original value, M_o. ΔE is the difference between the experimental measurement, E_p, and its original value, E_o. A positive value of the relative difference indicates that the model over-predicted the severity of the fire; for example, a higher temperature, lower oxygen concentration, higher smoke concentration, *etc.*

Finally, all of the calculations performed in the evaluation were *open*; that is, the heat release rate of the fire was a *specified* model input, and the results of the experiments were provided to the analysts.

A.1 Hot Gas Layer Temperature and Height

Relative differences for hot gas layer temperature were calculated using experimental data from ICFMP Benchmark Exercises (BE) #2, #3, #4, and #5; the FM/SNL test series; and the NBS multi-compartment fire test series. In the case of hot gas layer temperature, positive relative differences are an indication that MAGIC's predictions are higher than the experimental observations. By contrast, in the case of hot gas layer height, positive relative differences suggest that MAGIC's predictions are lower than the measured height.

A.1.1 ICFMP BE #2

The HGL temperature and depth were calculated from the averaged gas temperatures from three vertical thermocouple arrays using the standard reduction method. There were 10 thermocouples in each vertical array, spaced 2 m (6.6 ft) apart in the lower two-thirds of the hall, and 1 m (3.3 ft) apart near the ceiling. Figure A-1 presents a snapshot from one of the simulations.

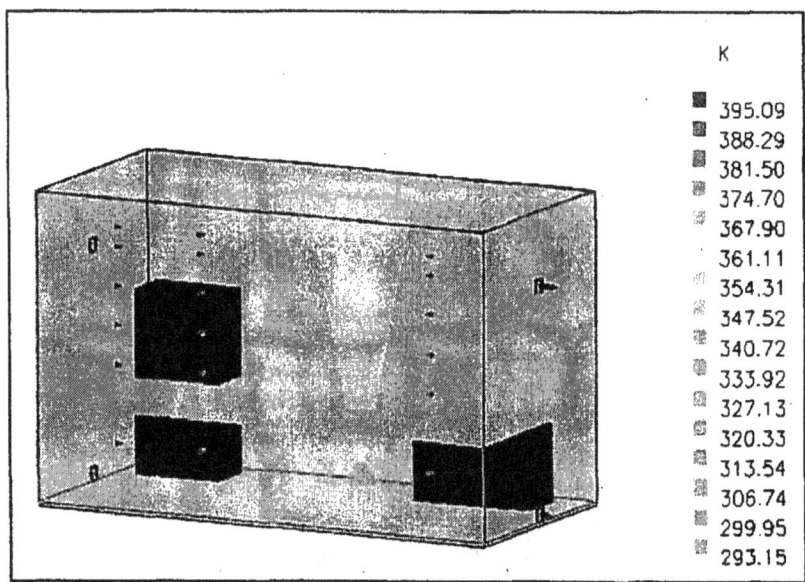

Figure A-1: Cut-Away View of the MAGIC Simulation of ICFMP BE #2, Case 2

The comparison between measured hot gas layer temperatures and heights for ICFMP BE #2, Cases 1, 2 and 3 are presented in Figure A-2.

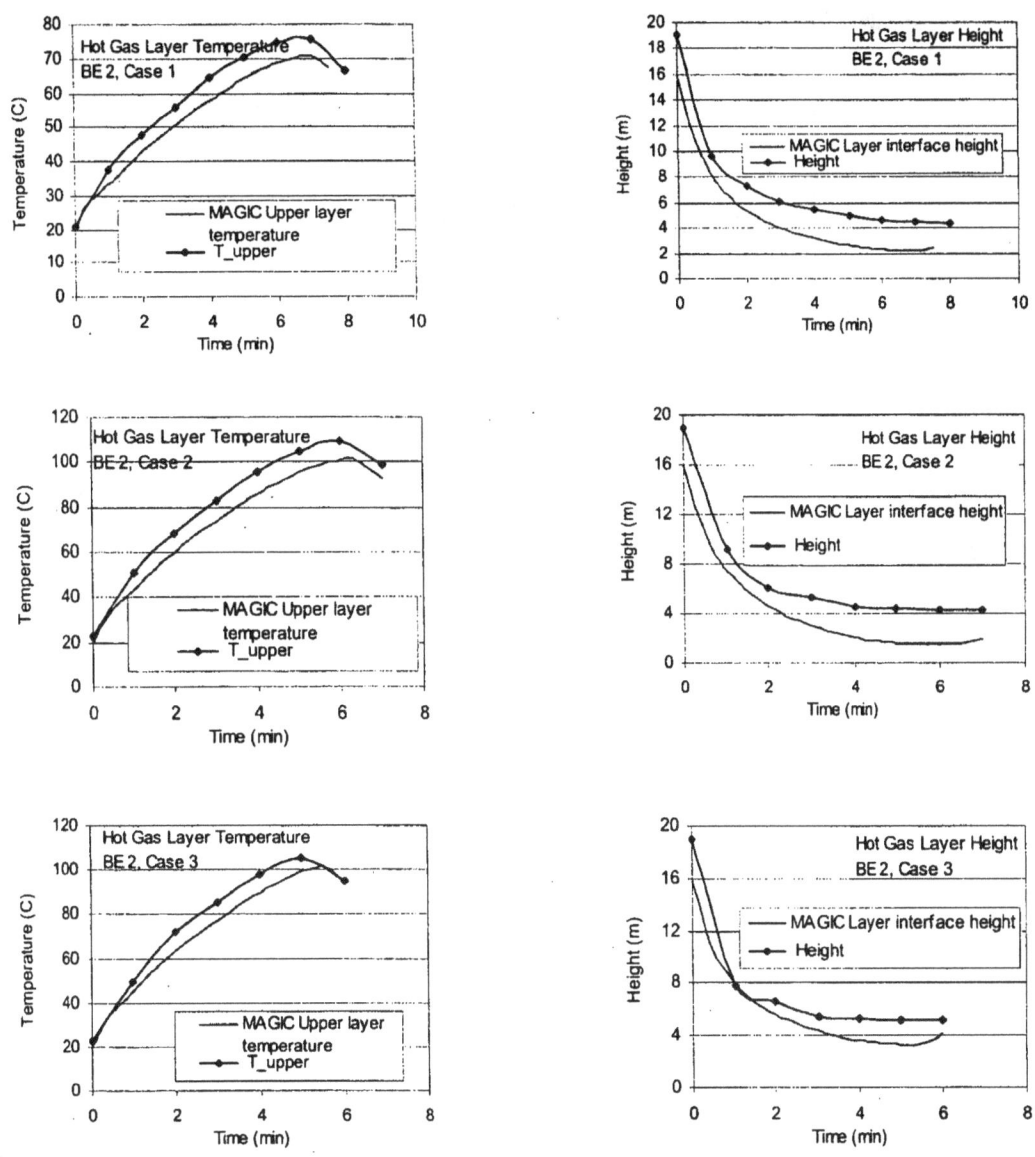

Figure A-2: Hot Gas Layer (HGL) Temperature and Height, ICFMP BE #2

Table A-1 summarizes the relative differences calculated for the hot gas layer temperature and height. MAGIC slightly under-predicts the temperature. At the same time, measured hot gas layer heights are consistently higher than the MAGIC prediction by a relatively small margin.

Table A-1: Relative Differences of Hot Gas Layer Temperature and Height in ICFMP BE#2

Test	Hot Gas Layer Temperature			Hot Gas Layer Height		
	ΔE (°C)	ΔM (°C)	Relative Difference	ΔE (m)	ΔM (m)	Relative Difference
ICFMP 2-1	54.8	50.8	-7%	-14.57	-13.66	-6%
ICFMP 2-2	86.3	81.6	-5%	-14.77	-14.35	-3%
ICFMP 2-3	82.6	81.3	-2%	-13.86	-12.55	-9%

A.1.2 ICFMP BE #3

BE #3 consisted of 15 liquid spray fire tests with different heat release rates, pan locations, and ventilation conditions. The basic geometry as modeled in MAGIC is shown in Figure A-3. Gas temperatures were measured using seven floor-to-ceiling thermocouple arrays (or "trees") distributed throughout the compartment. The average hot gas layer temperature and height were calculated using thermocouple Trees 1, 2, 3, 5, 6 and 7. Tree 4 was not used because one of its thermocouples (4-9) malfunctioned during most of the experiments.

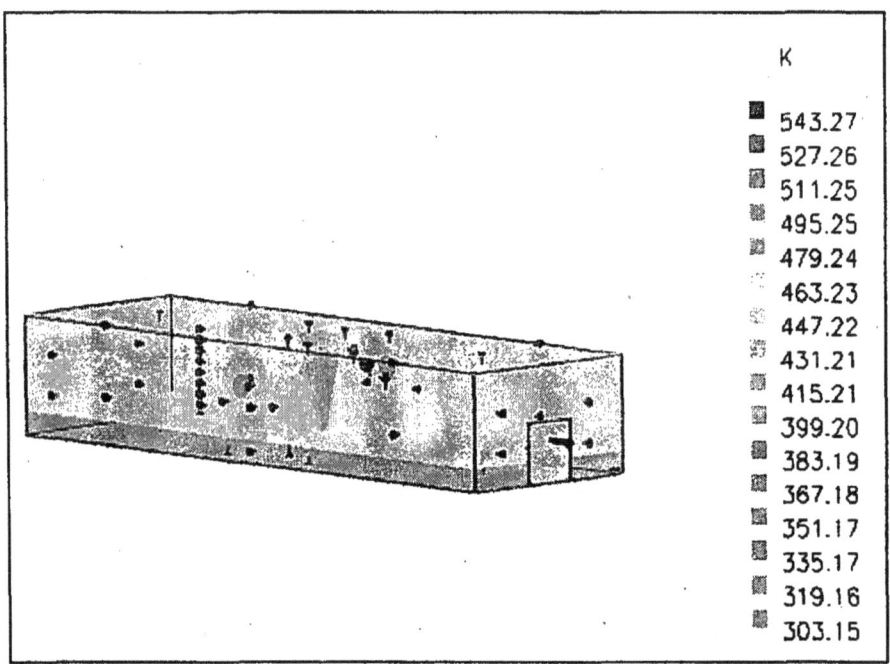

Figure A-3: Snapshot of the MAGIC Simulation of ICFMP BE #3, Test 3

In the closed-door tests, the HGL descended all the way to the floor. However, the reduction method, used on both the measured and predicted temperatures, does not account for the formation of a single layer and, therefore, does not indicate that the layer dropped all the way to the floor. Notice that in the MAGIC simulations, the hot gas layer is predicted to reach the floor.

It is important to indicate also that the HGL reduction method produces spurious results in the first few minutes of each test because no clear layer has yet formed.

The comparisons between MAGIC simulations and measured hot gas layer temperatures and heights are shown in Figure A-4 through Figure A-7.

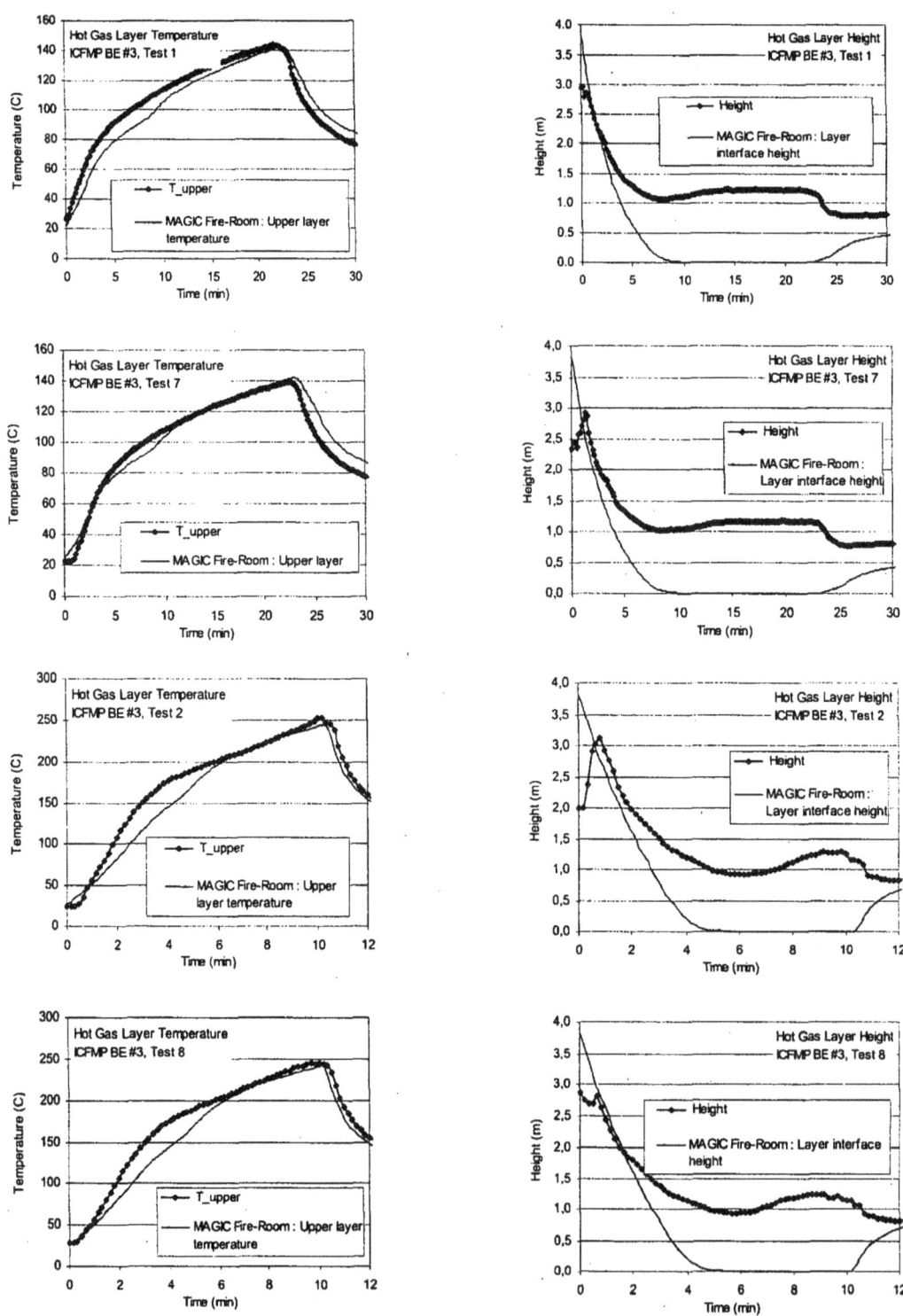

Figure A-4: Hot Gas Layer (HGL) Temperature and Height, ICFMP BE #3, Closed-Door Tests

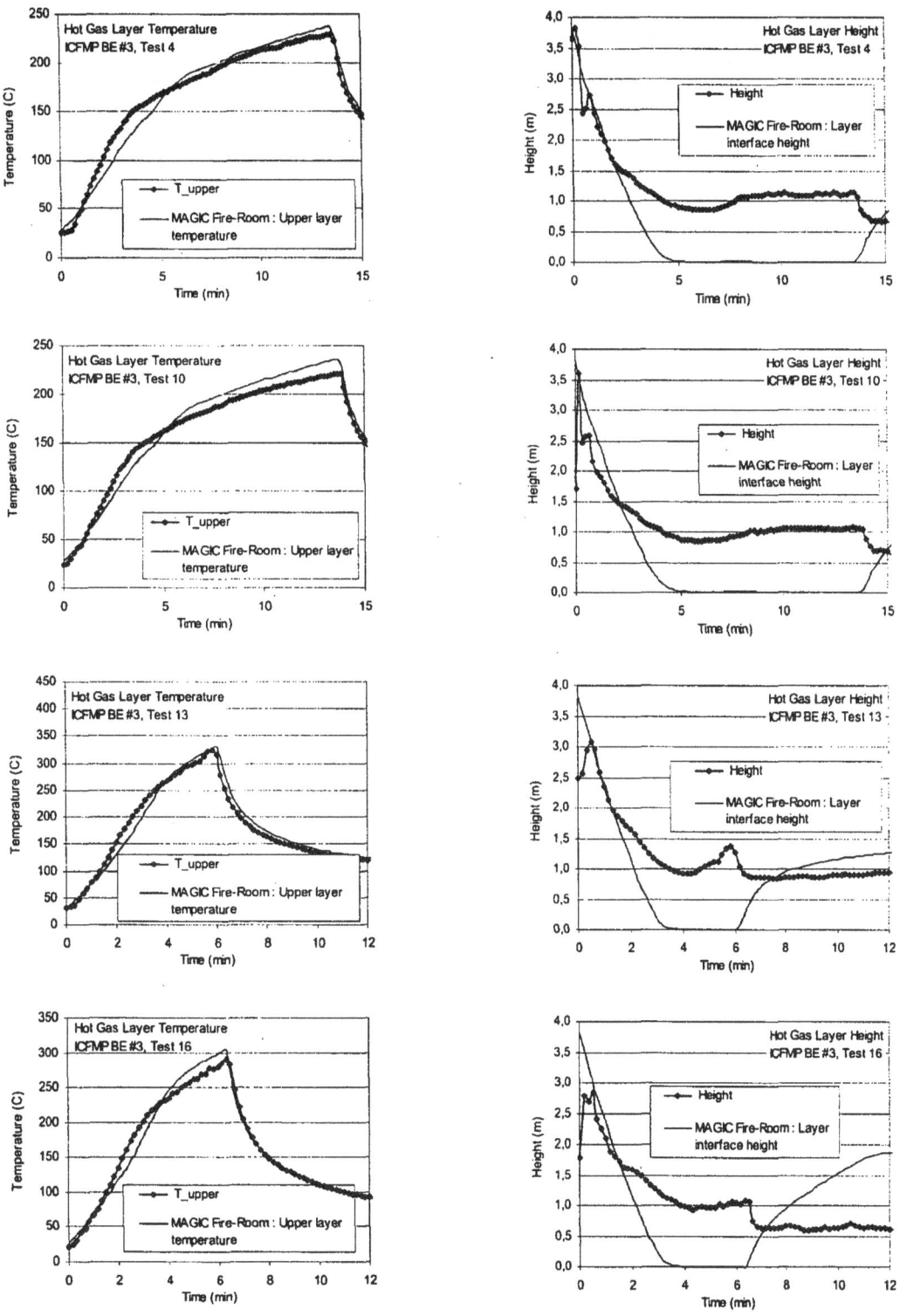

Figure A-5: Hot Gas Layer (HGL) Temperature and Height, ICFMP BE #3, Closed-Door Tests

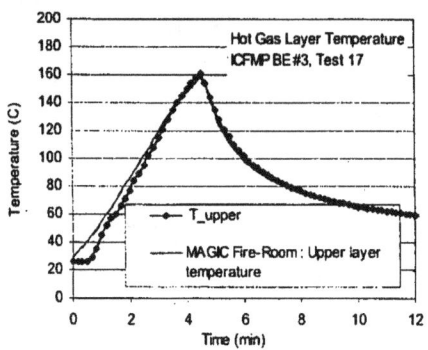
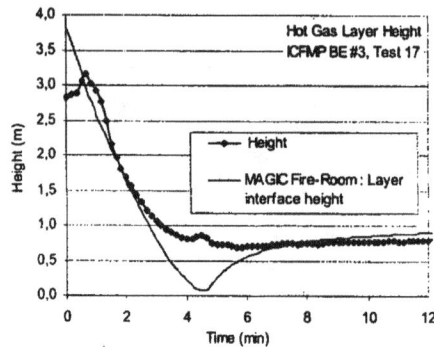

Open-door tests

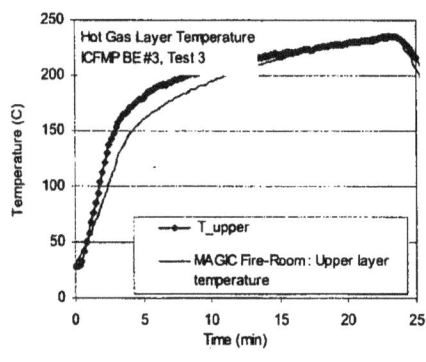
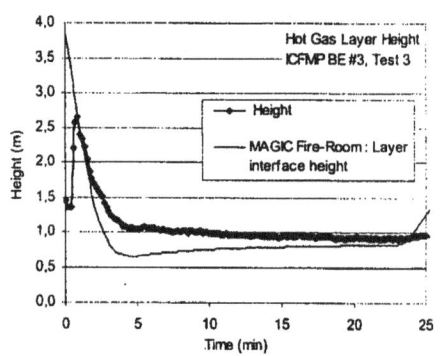

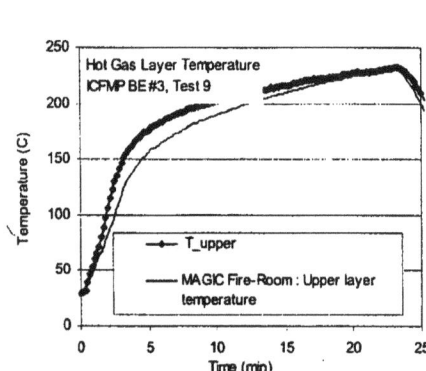
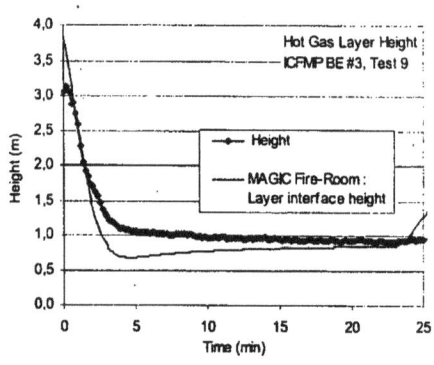

Figure A-6: Hot Gas Layer (HGL) Temperature and Height, ICFMP BE #3, Open-Door Tests

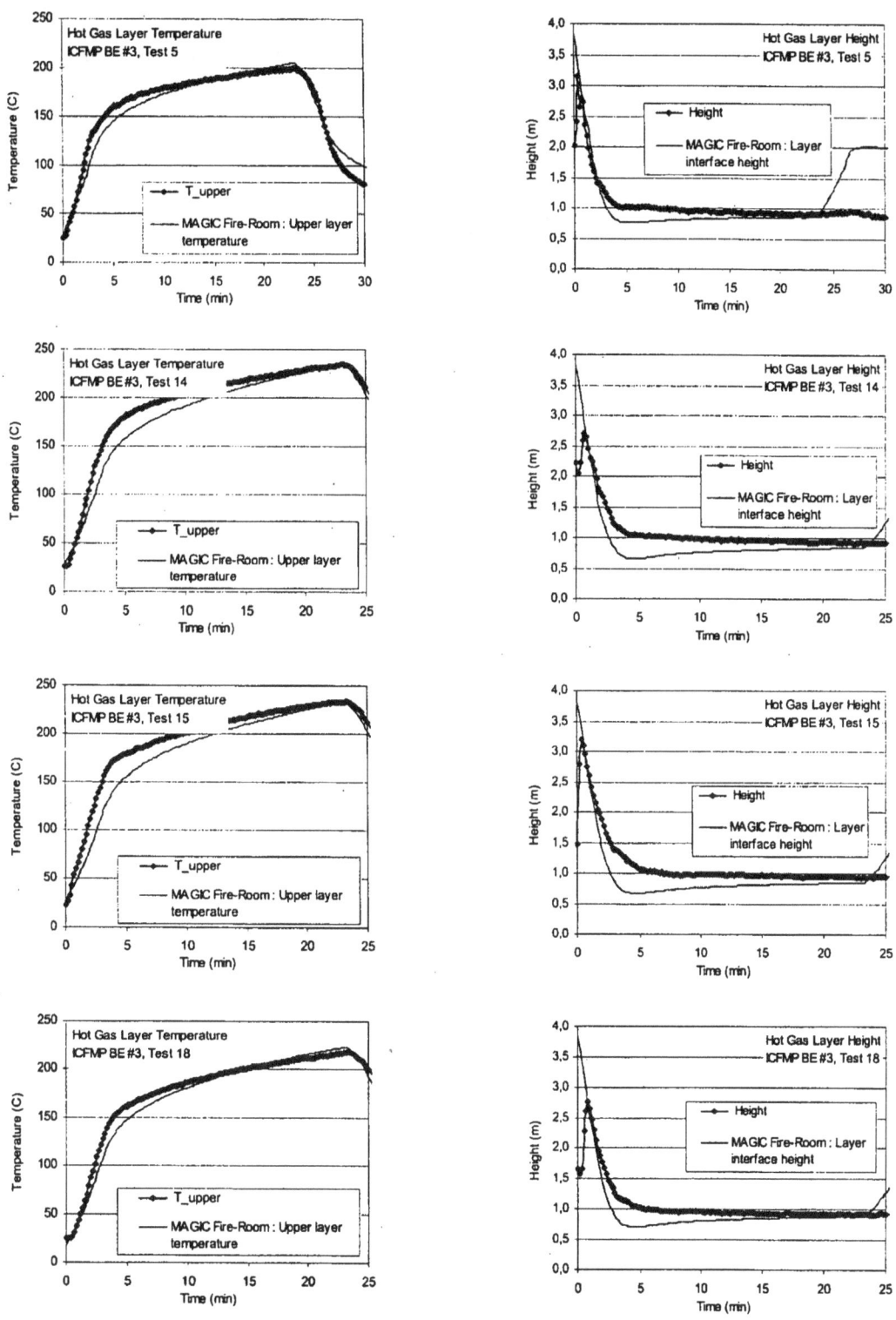

Figure A-7: Hot Gas Layer (HGL) Temperature and Height, ICFMP BE #3, Open-Door Tests

Table A-2: Relative Differences of Hot Gas Layer Temperature and Height in ICFMP BE#3

Test	Hot Gas Layer Temperature			Hot Gas Layer Height		
	ΔE (°C)	ΔM (°C)	Relative Difference	ΔE (m)	ΔM (m)	Relative Difference
ICFMP 3-1	122.9	120.3	-2%			N/A[1]
ICFMP 3-7	116.8	117.3	0%			N/A[1]
ICFMP 3-2	229.2	219.2	-4%			N/A[1]
ICFMP 3-8	217.7	218.3	0%			N/A[1]
ICFMP 3-4	204.3	210.8	3%			N/A[1]
ICFMP 3-10	197.8	209.4	6%			N/A[1]
ICFMP 3-13	290.5	298.9	3%			N/A[1]
ICFMP 3-16	268.4	278.6	4%			N/A[1]
ICFMP 3-17	135.3	129.2	-5%			N/A[1]
ICFMP 3-3	207.3	207.1	0%	-3.26	-3.82	17%
ICFMP 3-9	204.0	204.5	0%	-3.23	-3.82	18%
ICFMP 3-5	175.5	176.5	1%	-2.98	-3.82	28%
ICFMP 3-14	208.2	205.8	-1%	-3.29	-3.82	16%
ICFMP 3-15	210.6	205.3	-3%	-3.13	-3.75	20%
ICFMP 3-18	193.4	204.6	6%	-3.26	-3.82	17%

1. Relative difference not applicable for closed door compartment fire experiments because the data reduction method does not account for the formation of a single layer, which is the case when the hot gas layer reaches the floor.

A.1.3 ICFMP BE #4

ICFMP BE #4 consisted of two experiments, of which one (Test 1) was chosen for validation. Compared to the other experiments, this fire was relatively large in a relatively small compartment. Thus, its HGL temperature is considerably higher than the other fire tests under study. As shown in Figure A-8, the compartment geometry is fairly simple, consisting of a rectangular shaped room.

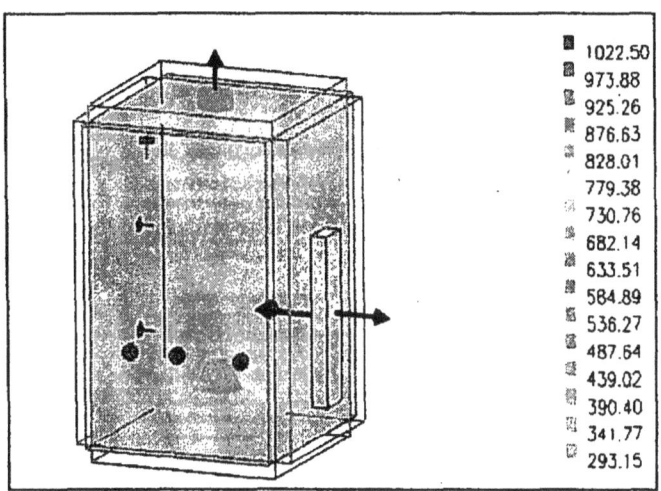

Figure A-8: Snapshot of the MAGIC Simulation of ICFMP BE #4, Test 1

Figure A-9 includes the comparison between experimental and predicted hot gas layer temperature and height. The relative differences calculated for this experiment are listed in Table A-3.

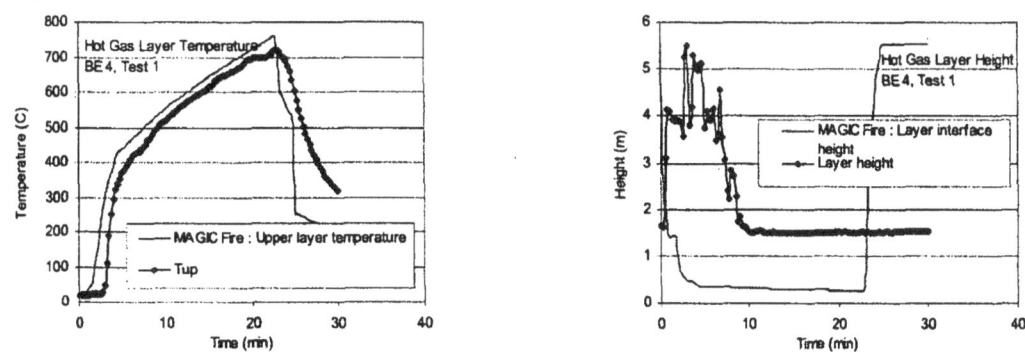

Figure A-9: Hot Gas Layer (HGL) Temperature and Height, ICFMP BE #4, Test 1

Table A-3: Relative Differences of Hot Gas Layer Temperature and Height in ICFMP BE#4

Test	Hot Gas Layer Temperature			Hot Gas Layer Height		
	ΔE (°C)	ΔM (°C)	Relative Difference	ΔE (m)	ΔM (m)	Relative Difference
ICFMP 4-1	700.1	741.4	6%	-4.20	-5.27	25%

A.1.4 ICFMP BE #5

BE #5 was performed in the same fire test facility as BE #4. Figure A-10 displays the overall geometry of the compartment, as idealized by MAGIC. Only one experiment (test 4) from this test series was used in the evaluation, and only the first 20 minutes of the test, during the "pre-heating" stage, when only the ethanol pool fire was active. The burner was lit after that point, and the cables began to burn.

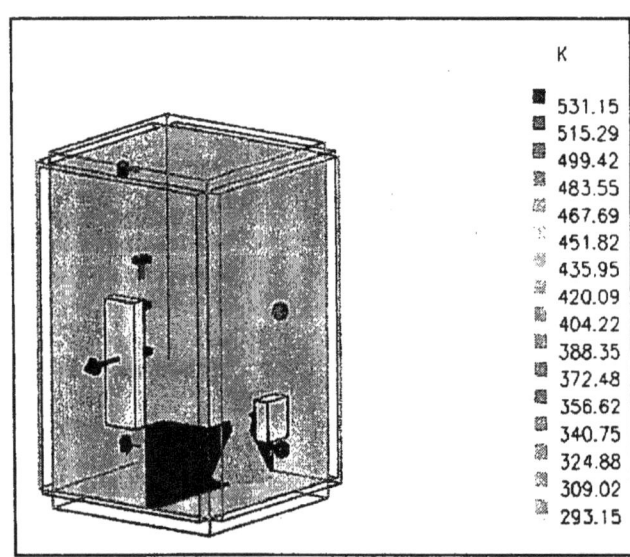

Figure A-10: Snapshot of the MAGIC Simulation of ICFMP BE #5, Test 4

Figure A-11 summarizes the comparison between the experimental and predicted hot gas layer and height during the first 20 minutes of the simulation. The corresponding relative differences are listed in Table A-4.

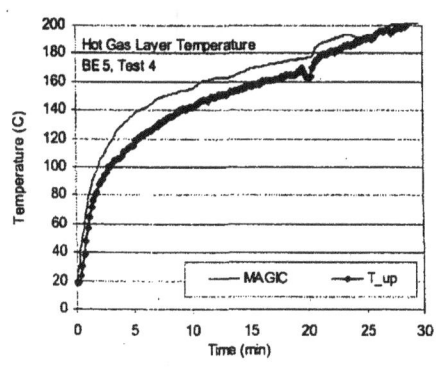 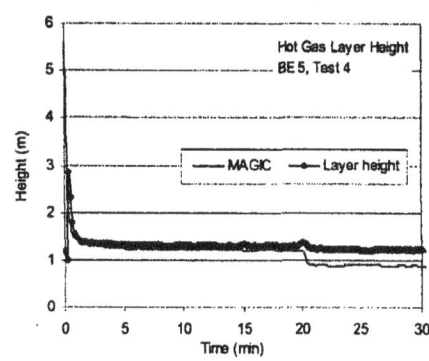

Figure A-11: Hot Gas Layer (HGL) Temperature and Height, ICFMP BE #5, Test 4

Table A-4: Relative Differences of Hot Gas Layer Temperature and Height in ICFMP BE#5

Test	Hot Gas Layer Temperature			Hot Gas Layer Height		
	ΔE (°C)	ΔM (°C)	Relative Difference	ΔE (m)	ΔM (m)	Relative Difference
ICFMP 5-4	185.7	182.8	-2%	-4.70	-4.72	1%

A.1.5 FM/SNL Test Series

Tests 4, 5, and 21 from the FM/SNL test series were selected for comparison. Figure A-12 provides a pictorial representation of the experimental geometry as idealized in MAGIC. The experimental hot gas layer temperature and height were calculated using the standard method. The thermocouple arrays that are referred to as Sectors 1, 2 and 3 were averaged (with an equal weighting for each) for Tests 4 and 5. For Test 21, only Sectors 1 and 3 were used, as Sector 2 fell within the smoke plume.

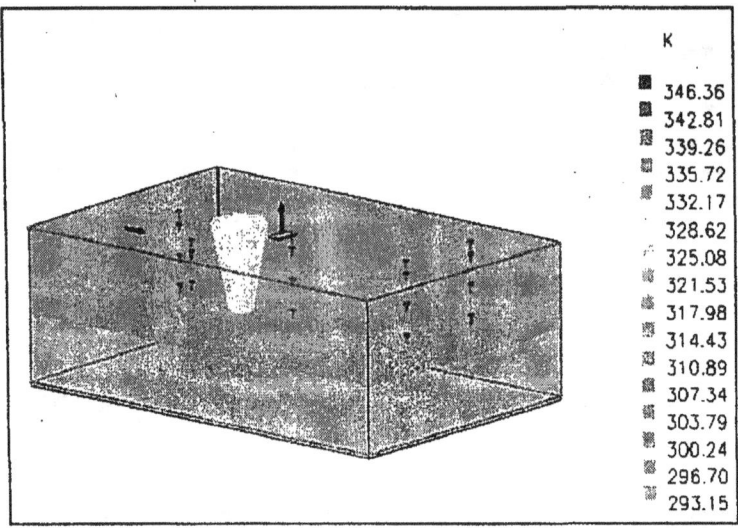

K
346.36
342.81
339.26
335.72
332.17
328.62
325.08
321.53
317.98
314.43
310.89
307.34
303.79
300.24
296.70
293.15

Figure A-12: Snapshot from the MAGIC Simulation of FM/SNL Test 5

Figure A-13 summarizes the graphical comparison of hot gas layer temperatures and heights for Tests 4, 5, and 21. The relative differences are included in Table A-5.

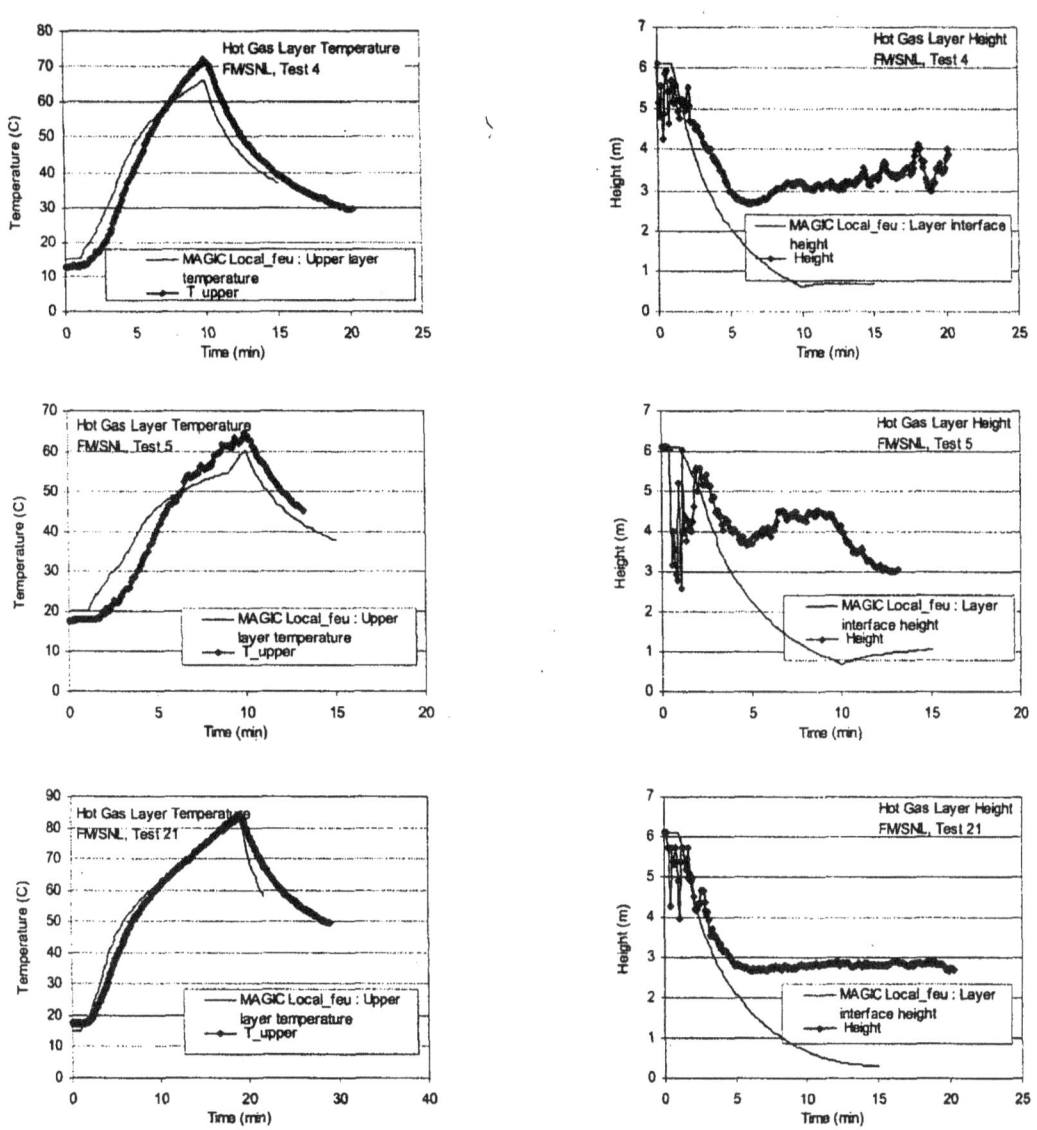

Figure A-13: Hot Gas Layer (HGL) Temperature and Height, FM/SNL Series

Table A-5: Relative Differences of Hot Gas Layer Temperature and Height in FM/SNL Series

Test	Hot Gas Layer Temperature			Hot Gas Layer Height		
	ΔE (°C)	ΔM (°C)	Relative Difference	ΔE (m)	ΔM (m)	Relative Difference
FM/SNL 4	59.2	51.0	-14%	-3.40	-5.50	N/A
FM/SNL 5	46.6	40.4	-13%	-3.23	-5.41	N/A
FM/SNL 21	66.0	59.6	-10%	-3.43	-5.79	N/A

A.1.6 The NBS Multi-Room Test Series

This series of experiments consisted of two relatively small rooms connected by a long corridor. The fire was located in one of the rooms. Eight vertical arrays of thermocouples were positioned throughout the test space (one in the burn room, one near the door of the burn room, three in the corridor, one in the exit to the outside at the far end of the corridor, one near the door of the other or "target" room, and one inside the target room). Four of the eight arrays were selected for comparison with model prediction (the array in the burn room, the array in the middle of the corridor, the array at the far end of the corridor, and the array in the target room). In Tests 100A and 100O, the target room was closed, in which case, the array in the exit doorway was used.

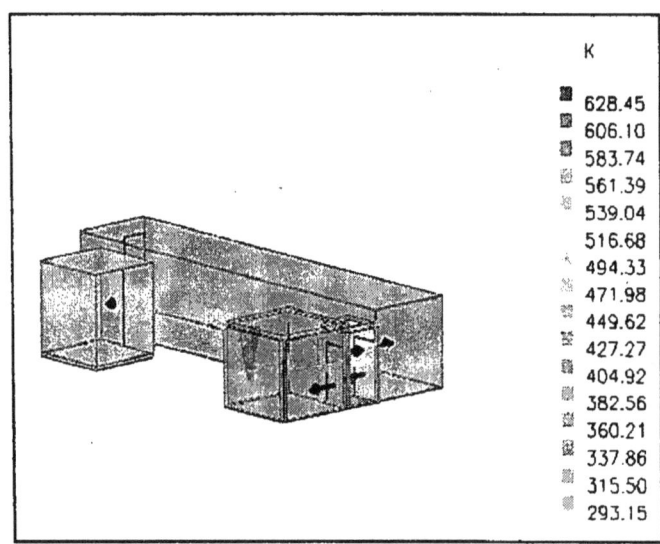

Figure A-14: Snapshot from the MAGIC Simulation of NBS Multi-Room Test 100Z

Figure A-15 through Figure A-17 compile the graphical comparison between experimental measurements and modeling results for hot gas layer temperature and height for the three selected experiments. Recall that the target room was closed in the first two experiments (A and O). Consequently, no relative difference was calculated for the target room in those two experiments. The relative differences are listed in Table A-6 through Table A-8.

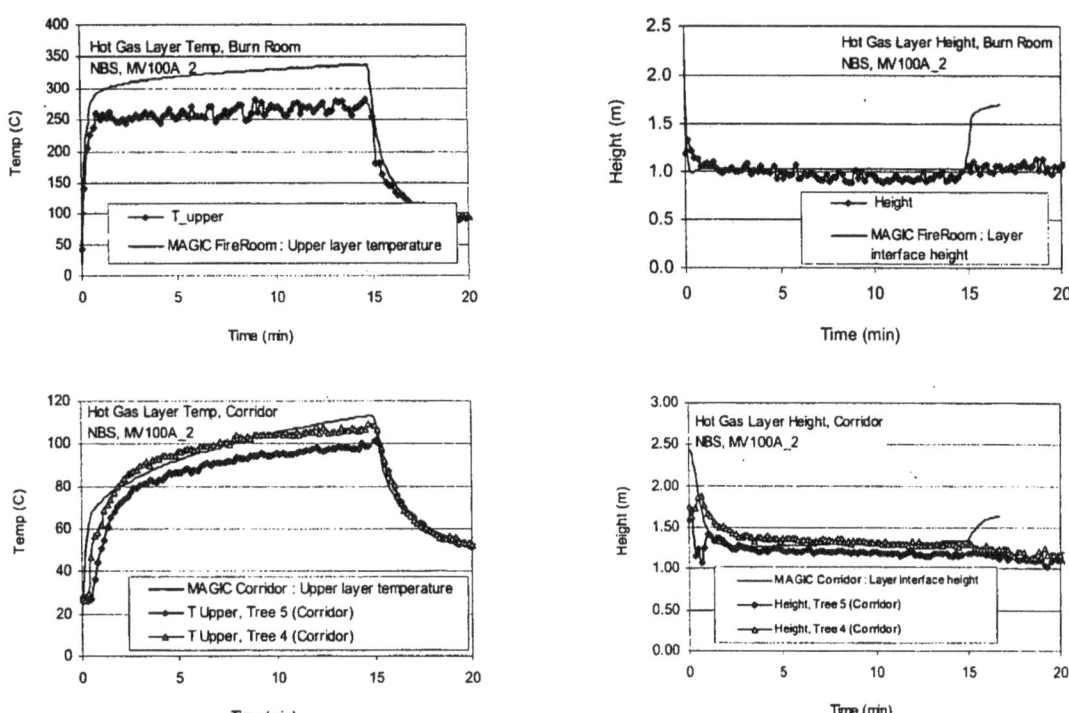

Figure A-15: Hot Gas Layer (HGL) Temperature and Height, NBS Multiroom Test 100A.

Table A-6: Relative Differences of Hot Gas Layer Temperature and Height in NBS Tests

		Hot Gas Layer Temperature			Hot Gas Layer Height		
		ΔE (°C)	ΔM (°C)	Relative Difference	ΔE (m)	ΔM (m)	Relative Difference
NBS A	Burn Room, Tree 1	239	318	33%	-1.27	-1.17	-8%
NBS A	Corridor, Tree 4	82	93	13%	-1.32	-1.18	-11%
NBS A	Corridor, Tree 5	75	93	25%	-1.41	-1.18	-16%
NBS A	Target Room, Tree 7	Target room closed					

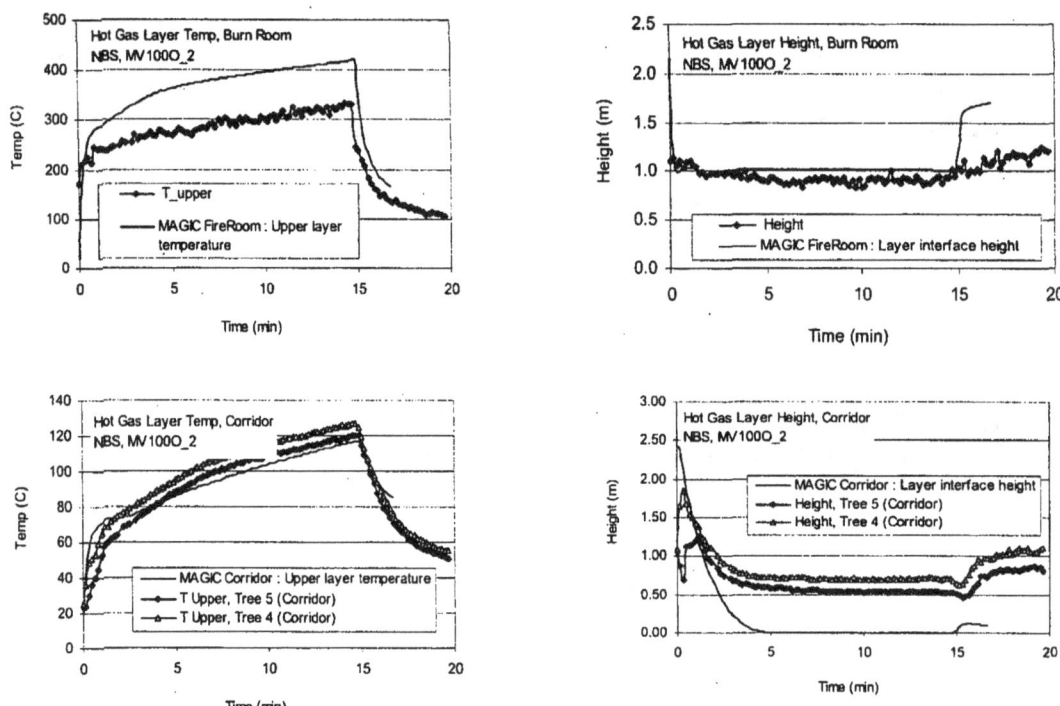

Figure A-16: Hot Gas Layer (HGL) Temperature and Height, NBS Multiroom Test 100O

Table A-7: Relative Differences of Hot Gas Layer Temperature and Height in NBS Tests

		Hot Gas Layer Temperature			Hot Gas Layer Height		
		ΔE (°C)	ΔM (°C)	Relative Difference	ΔE (m)	ΔM (m)	Relative Difference
NBS O	Burn Room, Tree 1	307	399	30%	-1.32	-1.19	-10%
NBS O	Corridor, Tree 4	102	97	-5%	-1.82	-2.44	23%
NBS O	Corridor, Tree 5	96	97	1%	-1.98	-2.44	23%
NBS O	Target Room, Tree 7	Target room closed					

A-18

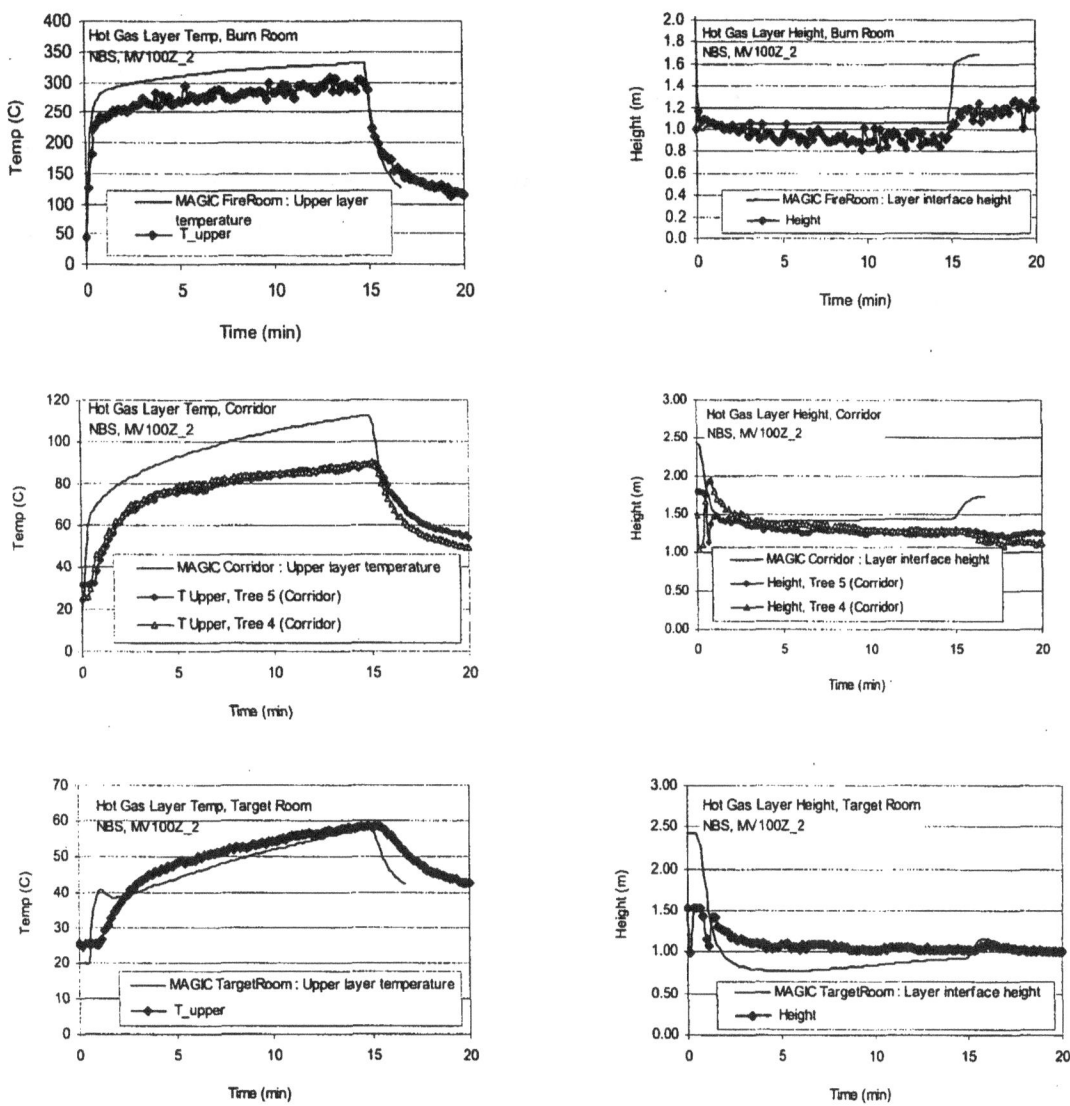

Figure A-17: Hot Gas Layer (HGL) Temperature and Height, NBS Multiroom Test 100Z

Table A-8: Relative Differences of Hot Gas Layer Temperature and Height in NBS Tests

		Hot Gas Layer Temperature			Hot Gas Layer Height		
		ΔE (°C)	ΔM (°C)	Relative Difference	ΔE (m)	ΔM (m)	Relative Difference
NBS Z	Burn Room, Tree 1	264	312	18%	-1.34	-1.17	-13%
NBS Z	Corridor, Tree 4	65	93	43%	-1.40	-1.05	-25%
NBS Z	Corridor, Tree 5	58	93	60%	-1.31	-1.05	-20%
NBS Z	Target Room, Tree 7	34	38	11%	-1.45	-1.66	14%

A.2 Ceiling Jet Temperature

MAGIC has an explicit ceiling jet temperature model based on the model developed by Cooper [Ref. 6]. The model also accounts for the hot gas layer effects using Cooper's method. In general, a target is specified in the computational domain. If the target is exposed to ceiling jet, the "Target/Gas Temperature" output option provides the ceiling jet temperature.

Experimental measurements for this category are available from ICFMP BE #3 and the FM/SNL series only.

Positive relative differences are an indication that the MAGIC prediction is higher than the experimental observation.

A.2.1 ICFMP BE #3

The thermocouple nearest the ceiling in Tree 7, located toward the back of the compartment, was chosen as a surrogate for the ceiling jet temperature. The 15 graphical comparisons of experimental measurements and model results are grouped in Figure A-18 and Figure A-19. The relative differences are listed in Table A-9.

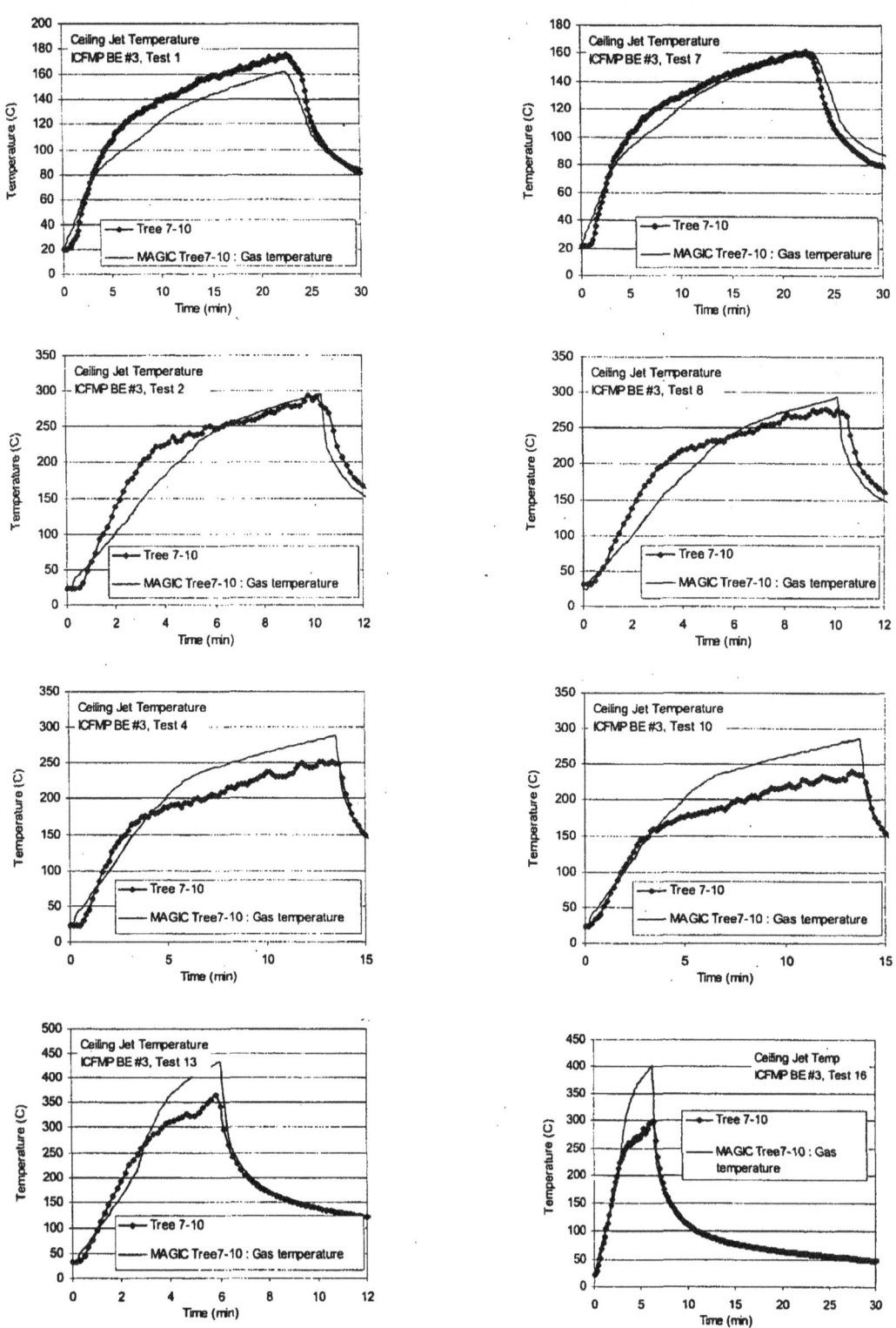

Figure A-18: Near-Ceiling Gas (Ceiling Jet) Temperatures, ICFMP BE #3, Closed-Door Tests

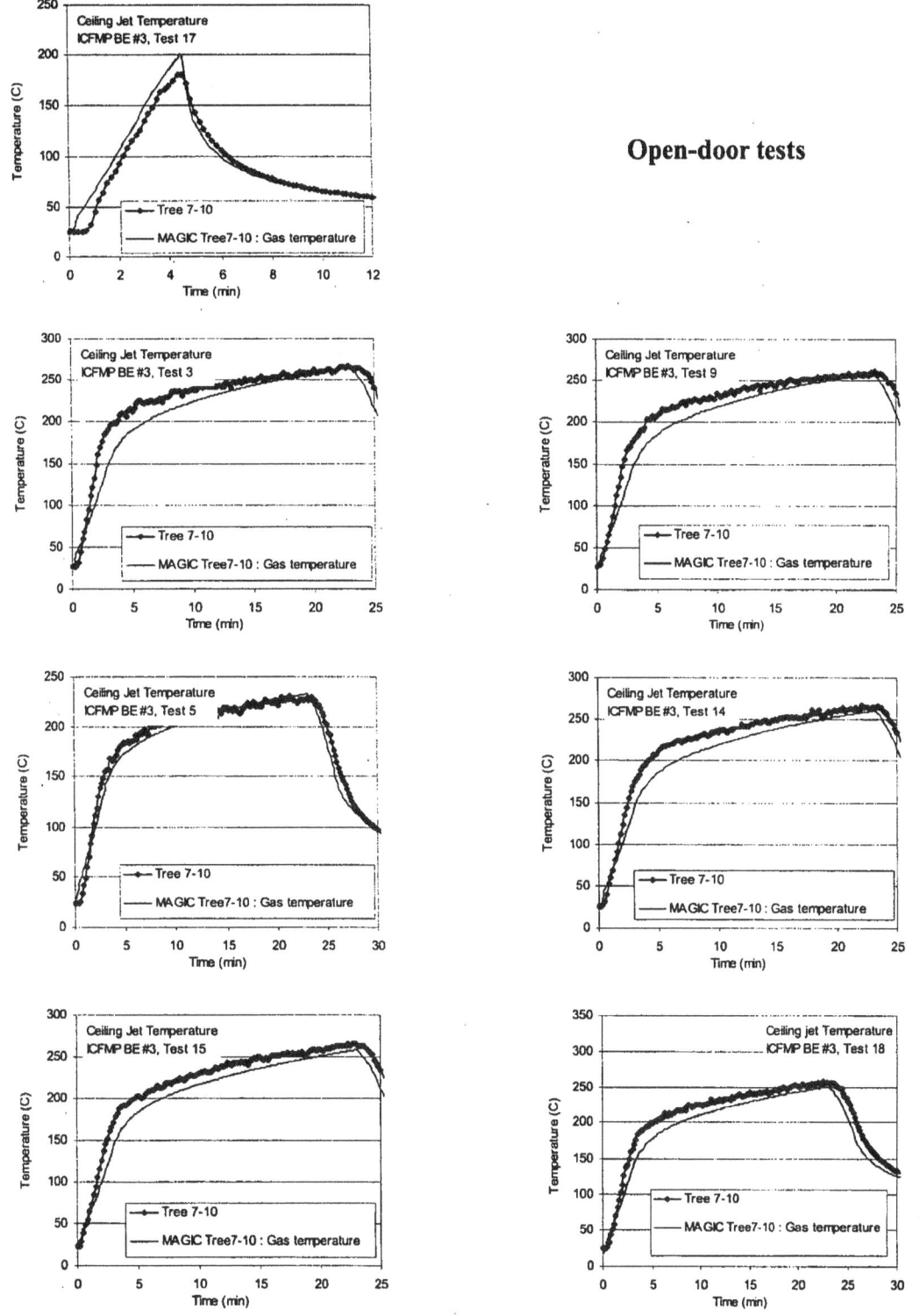

Figure A-19: Near-Ceiling Gas (Ceiling Jet) Temperatures, ICFMP BE #3, Open-Door Tests

Table A-9: Relative Differences for Ceiling Jet Temperature in ICFMP BE #3

	Instrument	ΔE (°C)	ΔM (°C)	Relative Difference
ICFMP 3-1	Tree 7-10	154.9	140.1	-10%
ICFMP 3-7	Tree 7-10	139.3	136.8	-2%
ICFMP 3-2	Tree 7-10	270.6	269.4	0%
ICFMP 3-8	Tree 7-10	246.9	268.3	9%
ICFMP 3-4	Tree 7-10	228.9	261.0	14%
ICFMP 3-10	Tree 7-10	217.5	259.1	19%
ICFMP 3-13	Tree 7-10	330.5	400.9	21%
ICFMP 3-16	Tree 7-10	277.7	376.6	36%
ICFMP 3-17	Tree 7-10	155.9	172.3	11%
ICFMP 3-3	Tree 7-10	240.7	234.6	-3%
ICFMP 3-9	Tree 7-10	234.6	231.4	-1%
ICFMP 3-5	Tree 7-10	207.7	204.7	-1%
ICFMP 3-14	Tree 7-10	240.8	232.5	-3%
ICFMP 3-15	Tree 7-10	243.7	231.4	-5%
ICFMP 3-18	Tree 7-10	235.1	233.2	-1%

A.2.2 The FM/SNL Test Series

The near-ceiling thermocouples in Sectors 1 and 3 were chosen as surrogates for the ceiling jet temperature. Figure A-20 compiles the graphical comparisons between experimental measurements for ceiling jet temperature and the MAGIC predictions. The corresponding relative differences are listed in Table A-10.

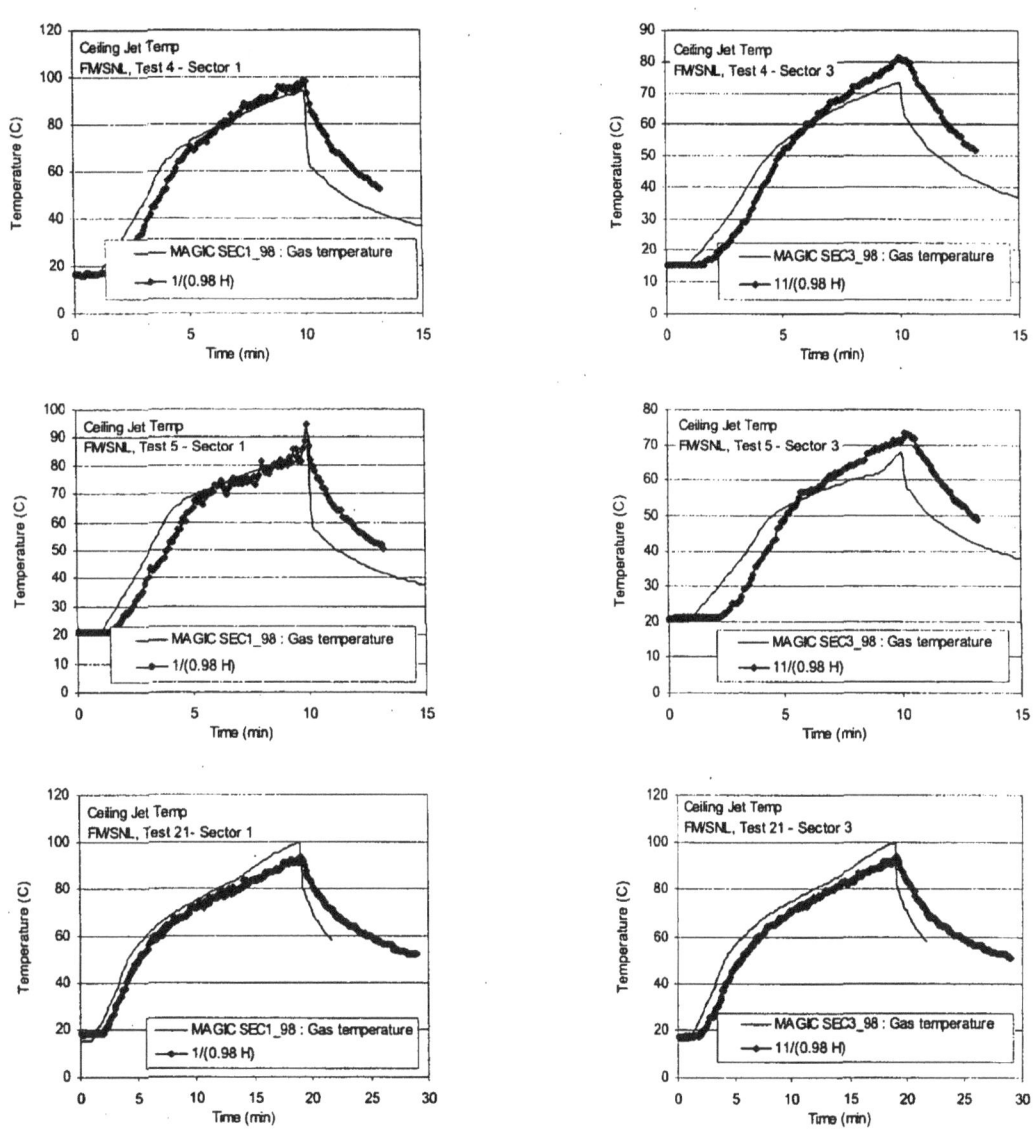

Figure A-20: Near-Ceiling Gas (Ceiling Jet) Temperatures, FM/SNL Series, Sectors 1 and 3

Table A-10: Relative Differences for Ceiling Jet Temperature in FM/SNL Tests

	Instrument	ΔE (°C)	ΔM (°C)	Relative Difference
FM/SNL 4	1/98H	82.8	78.8	-5%
	11/98H	66.1	58.1	-12%
FM/SNL 5	1/98H	73.7	66.8	-9%
	11/98H	52.6	47.5	-10%
FM/SNL 21	1/98H	75.9	74.2	-2%
	11/98H	77.2	74.2	-4%

A.3 Plume Temperature

Plume temperature measurements are available from ICFMP BE #2 and the FM/SNL series. For all other series of experiments, the temperature was not measured above the fire (BE #3), the fire plume leaned because of the flow pattern within the compartment (BE #4), or the fire was set up against a wall (NBS). Only for BE #2 and the FM/SNL series were the plumes relatively free from perturbations.

Once a target is specified in the calculation domain, MAGIC identifies if it is located within the fire plume region. Plume temperature results are captured in the "Target/Gas Temperature" output option. The output would include hot gas layer effects in the plume temperature if the target is located inside the fire plume and above the hot gas layer interface. Positive accuracies are an indication that MAGIC's predictions are higher than the experimental observations.

A.3.1 ICFMP BE #2

BE #2 consisted of liquid fuel pan fires conducted in the middle of a large fire test hall. Plume temperatures were measured at two heights above the fire, 6 m (19.7 ft) and 12 m (39.4 ft). The flames extended to about 4 m (13.1 ft) above the fire pan (Figure A-21). The suspended rectangle contains an array of thermocouples designed to locate the plume centerline. Notice that the smoke plume does not always rise straight up because of air currents within the large test hall.

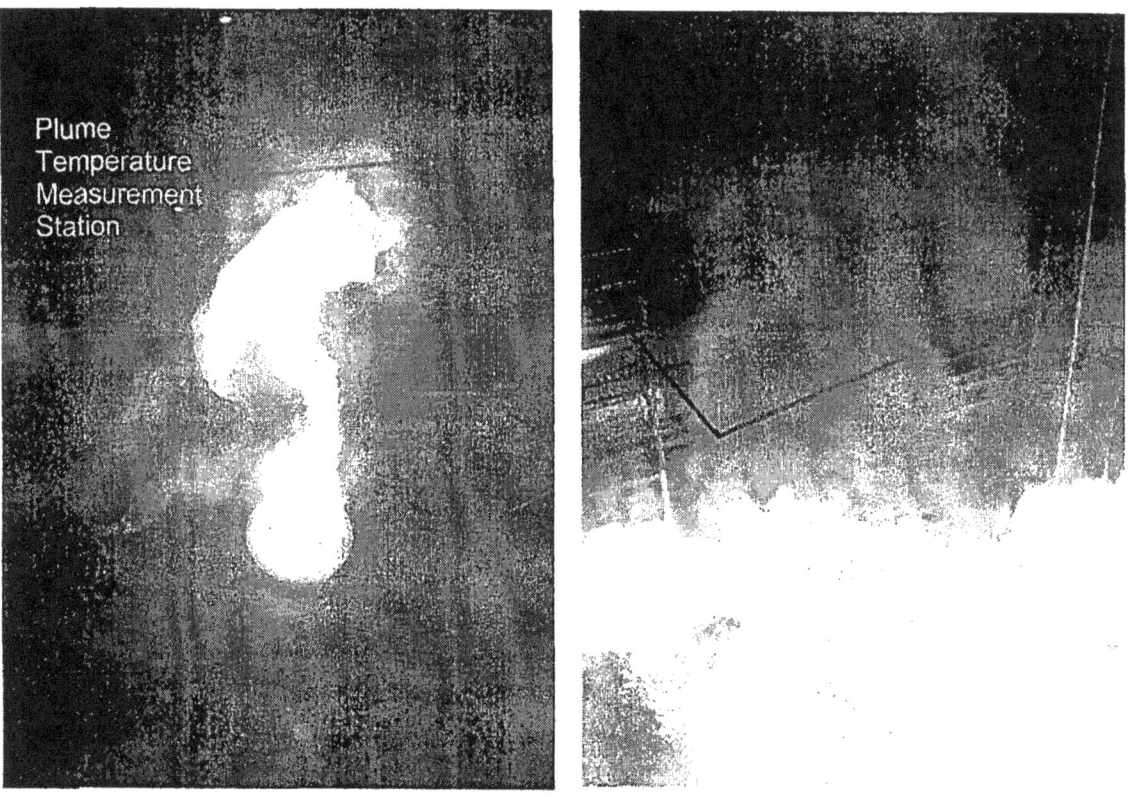

Figure A-21: Fire Plumes in ICFMP BE #2
(Courtesy of Simo Hostikka, VTT Building and Transport, Espoo, Finland)

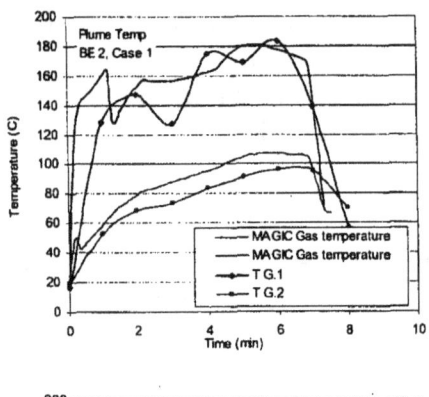
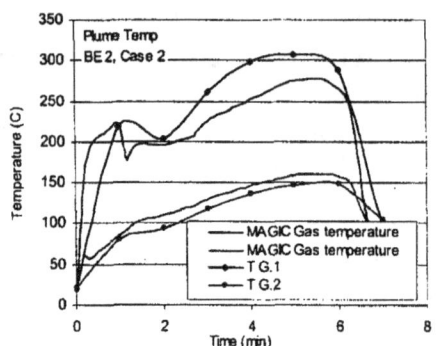
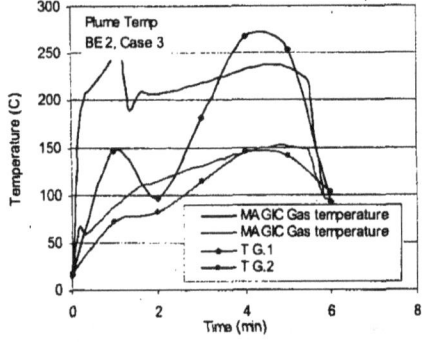

Figure A-22: Near-Ceiling Gas Temperatures, FM/SNL Series, Sectors 1 and 3

Figure A-22 illustrates the graphical comparison between experimental plume temperatures and MAGIC's predictions for the three cases in ICFMP BE #2. The corresponding relative differences are listed in Table A-11.

Table A-11: Relative Differences for Plume Temperature in ICFMP BE #2

	Instrument	ΔE (°C)	ΔM (°C)	Relative Difference
ICFMP 2-1	TG.1	166	161	-3%
	TG.2	79	87	10%
ICFMP 2-2	TG.1	288	258	-11%
	TG.2	128	141	10%
ICFMP 2-3	TG.1	252	229	-9%
	TG.2	128	132	3%

A.3.2 The FM/SNL Test Series

In Tests 4 and 5, thermocouples were positioned near the ceiling directly [5.9 m (19.4 ft)] over the fire pan. In Test 21, the fire pan was inside a cabinet. For that reason, no plume temperature comparison has been made. Figure A-23 presents the graphical comparisons and Table A-12 lists the corresponding relative differences.

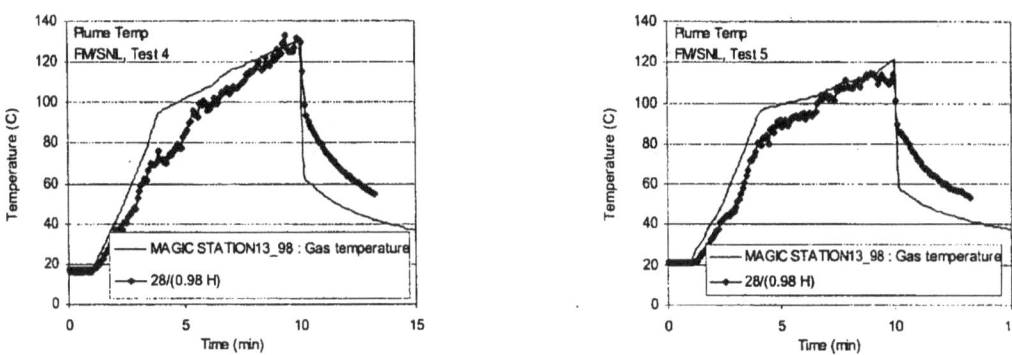

Figure A-23: Near-Plume Temperatures, FM/SNL Series, Sector 13

Table A-12: Relative Differences of Plume Temperature in FM/SNL Tests

	Instrument	ΔE (°C)	ΔM (°C)	Relative Difference
FM/SNL 4	28/98H	116	115	0%
FM/SNL 5	28/98H	94	101	8%

A.4 Flame Height

Flame height is recorded by visual observations, photographs, or video footage. Videos from the ICFMP BE #3 test series and photographs from BE #2 are available. It is difficult to precisely measure the flame height, but the photos and videos allow one to make estimates accurate to within a pan diameter.

A.4.1 ICFMP BE #2

Shown in Figure A-24 are MAGIC's predictions for flame height. Figure A-25 contains photographs of the actual fire. The height of the visible flame in the photographs has been estimated to be between 2.4 and 3 pan diameters [3.8 m to 4.8 m (12.5 to 15.7 ft)]. The height of the simulated fire fluctuates from 5 m (16.4 ft) to 6 m (19.7 ft) during the peak heat release rate phase.

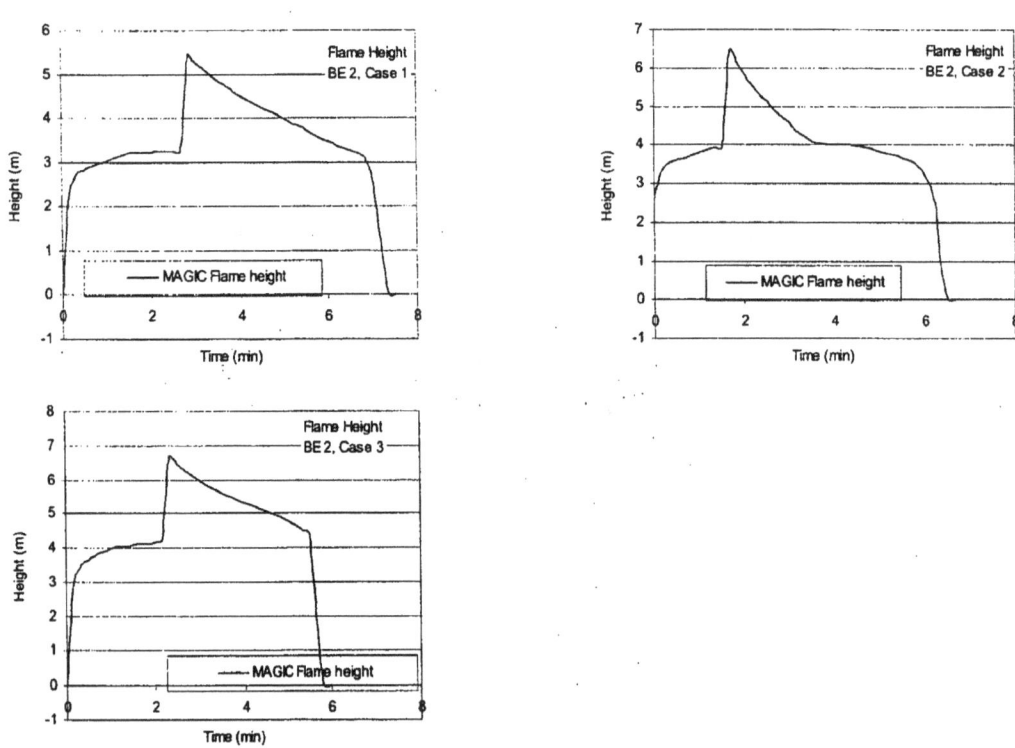

Figure A-24: Flame Heights for ICFMP BE #2

**Figure A-25: Photographs of Heptane Pan Fires, ICFMP BE #2, Case 2
(Courtesy of Simo Hostikka, VTT Building and Transport, Espoo, Finland)**

A.4.2 ICFMP BE #3

No measurements were made of the flame height during BE #3, but numerous photographs were taken. Figure A-26 is one of these photographs. These photographs provide at least a qualitative assessment of the MAGIC flame height prediction. Recall that the door is 2.0 m (6.6 ft) high. Inspection of the picture suggests that the flame height, at least is some of its oscillations, can be more than 2.0 m (6.6 ft) high. MAGIC, however, appears to over-predict flame heights, as most of the predictions are over 3 m (9.8 ft) high.

**Figure A-26: Photograph and Simulation of ICFMP BE #3, Test 3,
as seen through the 2 m x 2 m doorway (Photo courtesy of Francisco Joglar, SAIC)**

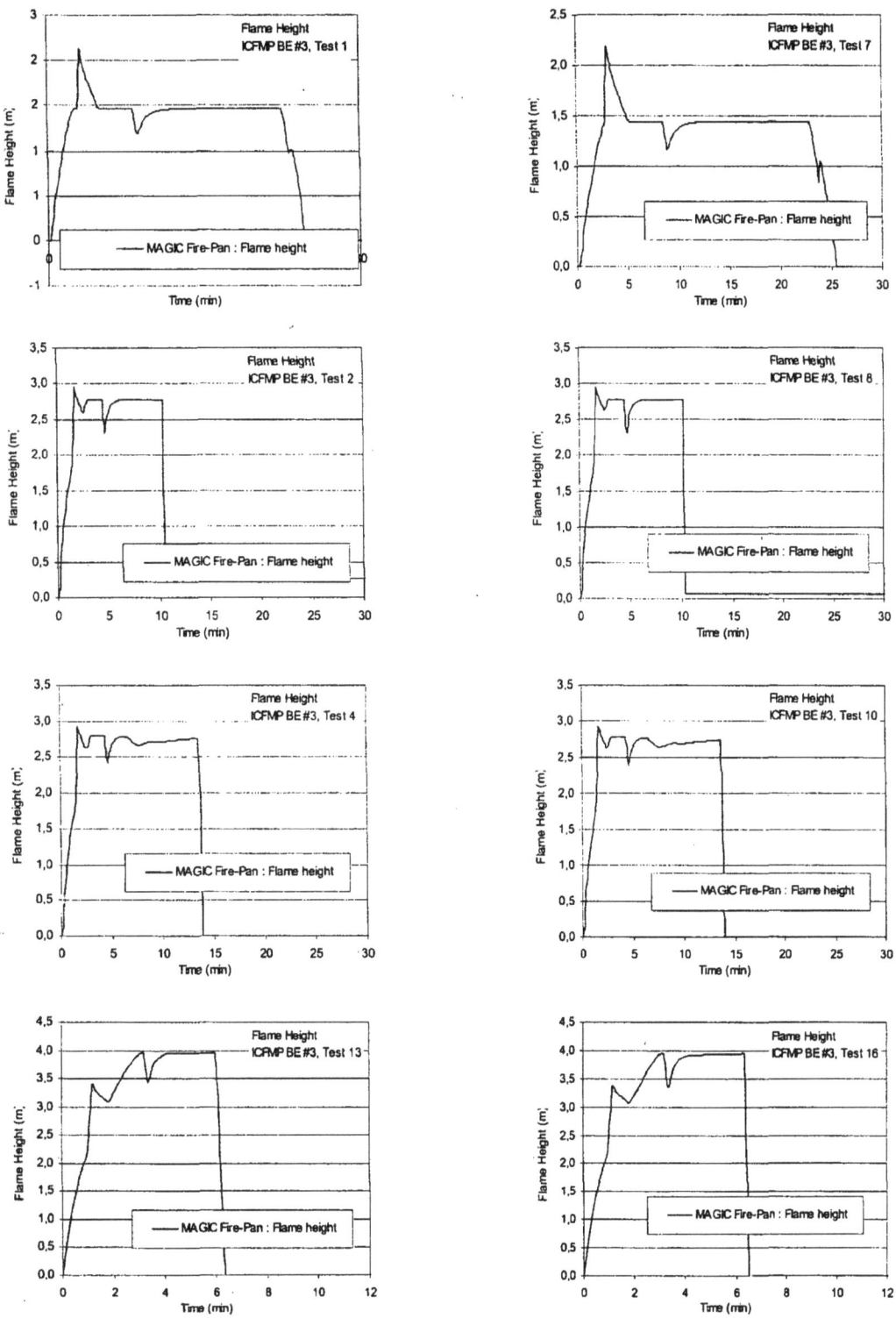

Figure A-27: Near-Ceiling Gas Temperatures, ICFMP BE #3, Closed-Door Tests

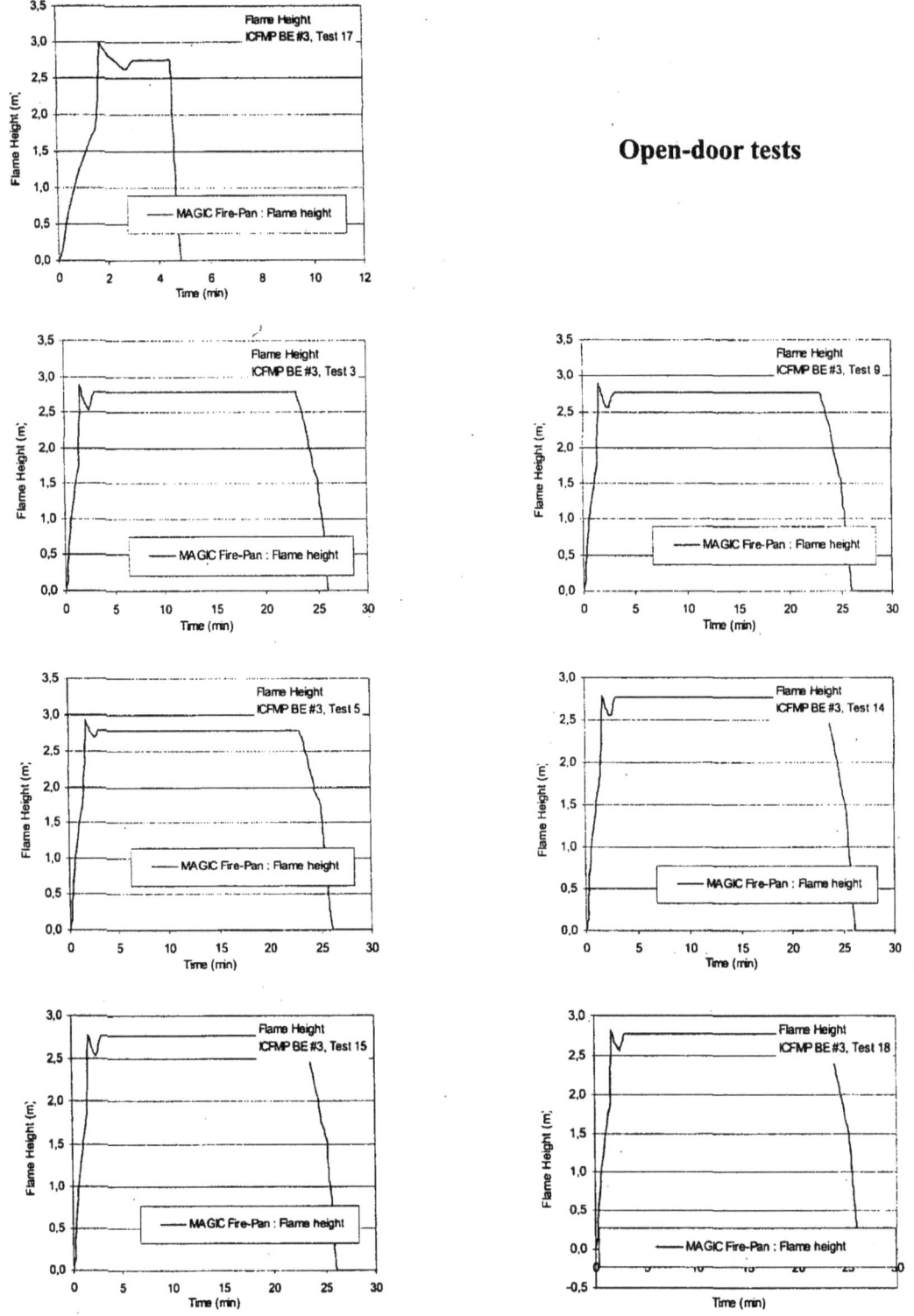

Figure A-28: Flame Heights, ICFMP BE #3, Open-Door Tests

A.5 Oxygen Concentration

Oxygen concentration data are available for accuracy calculations in ICFMP BE #3 and #5. For the calculations, measured values in the experiments are compared with the *upper-layer oxygen concentration*, which is an output available in MAGIC. Relative differences are calculated by comparing the lowest concentration measured in the experiments with the lowest concentration predicted by MAGIC. Positive relative differences indicate that MAGIC predicted a lower concentration than that measured in the experiments.

A.5.1 ICFMP BE #3

In experiments with closed room doors in the ICFMP BE #3 test series, the fuel supply was discontinued when the oxygen concentration was measured around 12 to 15%. In these tests, relative differences are measured at the lowest concentration before the experiment was terminated.

The graphical comparisons for closed- and open-door tests and relative differences are provided in Figure A-29, Figure A-30, and Table A-13, respectively.

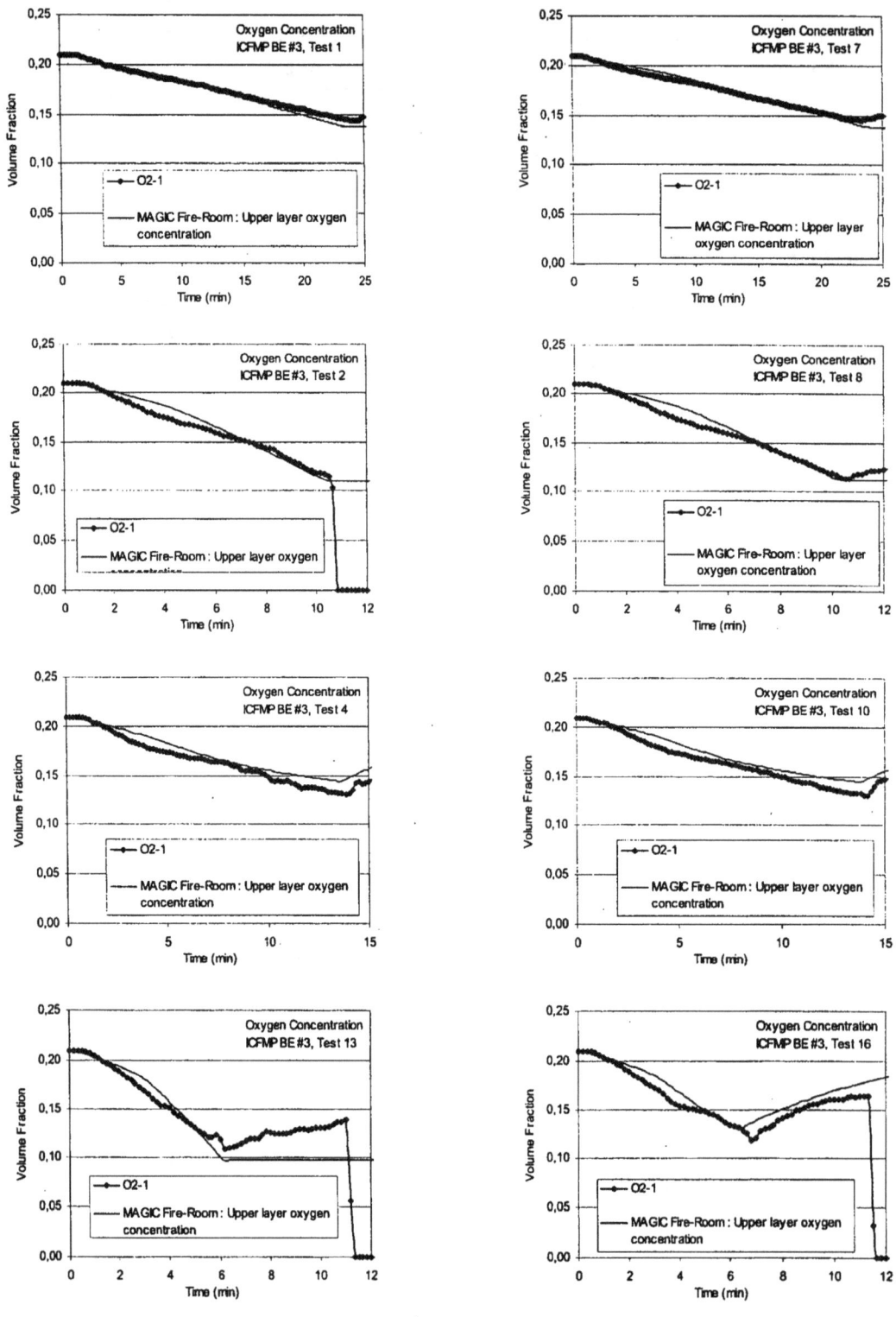

Figure A-29: Oxygen Concentration, ICFMP BE #3, Closed-Door Tests

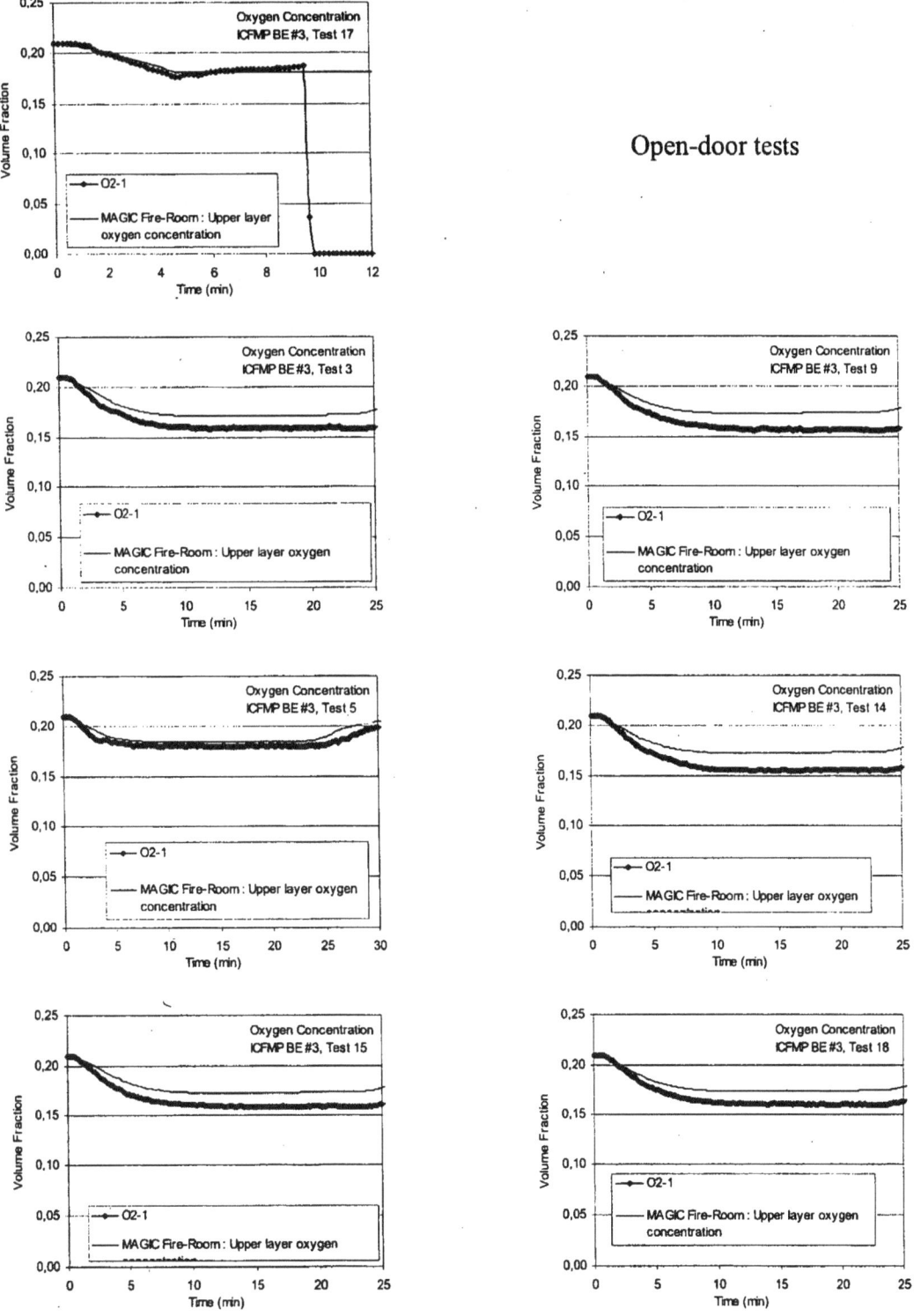

Figure A-30: Oxygen Concentration ICFMP BE #3, Open-Door Tests

Table A-13: Relative Differences for Oxygen Concentration in ICFMP BE #3 Tests

	ΔE	ΔM	Relative Difference
ICFMP 3-1	-0.065	-0.070	8%
ICFMP 3-7	-0.064	-0.068	6%
ICFMP 3-2	-0.092	-0.100	8%
ICFMP 3-8	-0.096	-0.098	2%
ICFMP 3-4	-0.079	-0.065	-17%
ICFMP 3-10	-0.079	-0.065	-18%
ICFMP 3-13	-0.101	-0.110	9%
ICFMP 3-16	-0.091	-0.081	-11%
ICFMP 3-17	-0.033	-0.028	-16%
ICFMP 3-3	-0.052	-0.039	-24%
ICFMP 3-9	-0.054	-0.038	-30%
ICFMP 3-5	-0.030	-0.025	-17%
ICFMP 3-14	-0.055	-0.038	-31%
ICFMP 3-15	-0.052	-0.038	-27%
ICFMP 3-18	-0.051	-0.037	-27%

A.5.2 ICFMP BE #5

Figure A-31 and Table A-14 present the graphical comparison and relative difference for oxygen concentration in ICFMP BE #5, Test 4.

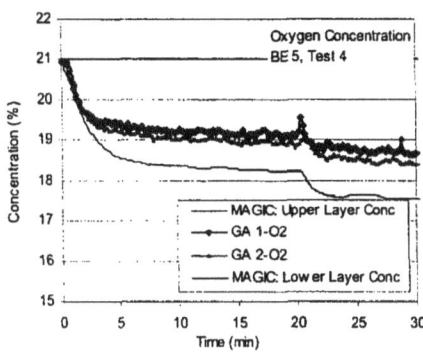

Figure A-31: Oxygen Concentration ICFMP BE #5, Test 4

Table A-14: Relative Differences for Oxygen Concentration in ICFMP BE #5 Tests

	ΔE	ΔM	Relative Difference
ICFMP 5-4	-0.028	-0.035	24%

A.6 Smoke Concentration

Data for smoke concentration are only available in ICFMP BE #3. Positive accuracies indicate that MAGIC predicted higher smoke concentrations than those measured in the experiments. Depending on the application, this may not yield a conservative result.

The units used for accuracy calculations are mg/m^3. Notice that MAGIC's output is average extinction coefficient, k, with units of 1/m. As a result, the direct output from the model was converted to mg/m^3 using the following equation:

$$v = k \Big/ k_m$$

where v is the concentration in mg/m^3, and k_m is a constant with value 0.0076 m^2/mg [Ref. 21].

A.6.1 ICFMP BE #3

Figure A-32 and Figure A-33 contain comparisons of measured and predicted smoke concentration at one measuring station in the upper layer for closed- and open-door tests. MAGIC consistently under-predicts the smoke concentration, with the exception of Test 17, which consisted of a Toluene fuel. This trend is reflected in the relative differences listed in Table A-15.

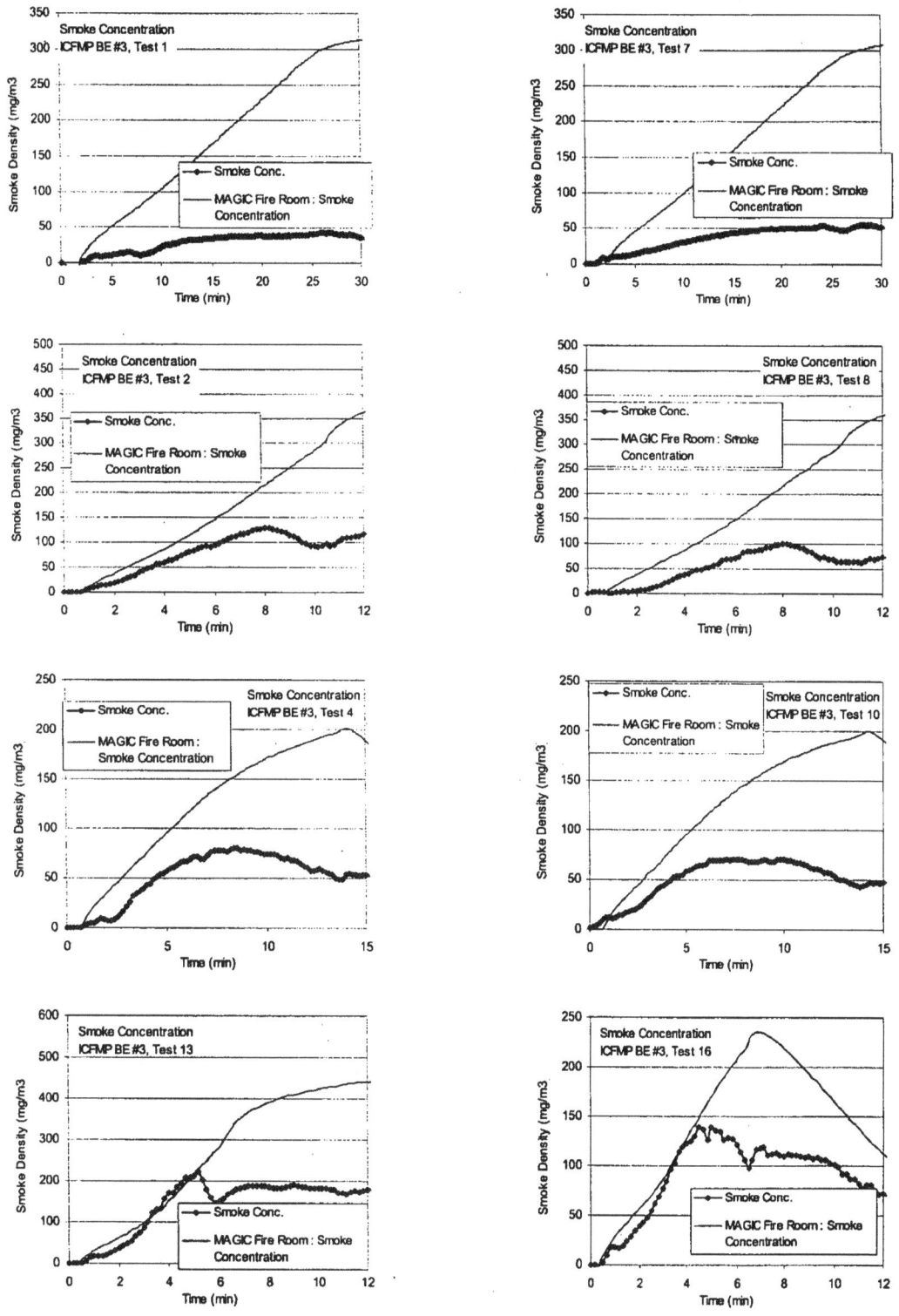

Figure A-32: Smoke Concentration in ICFMP BE #3, Closed-Door Tests

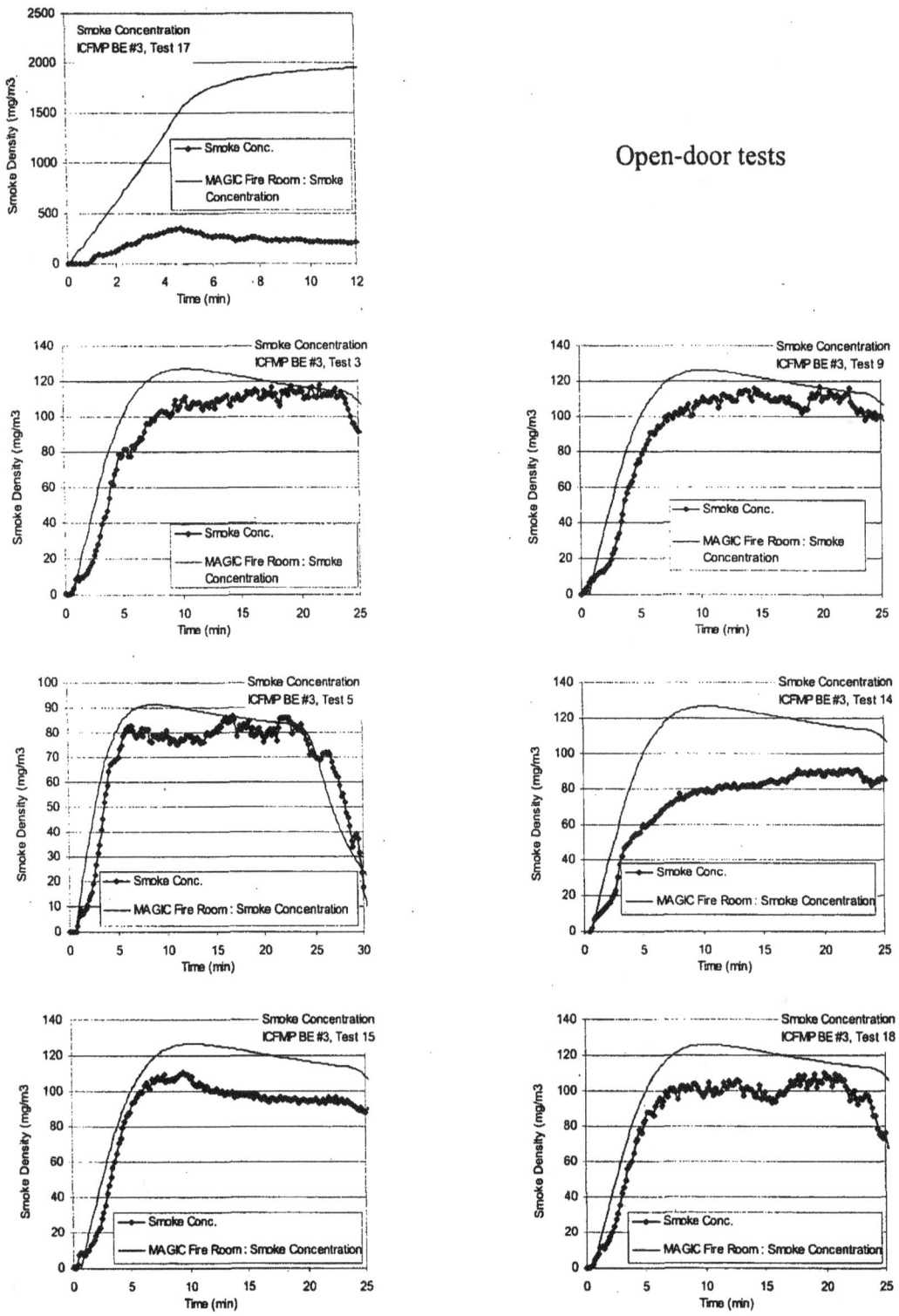

Figure A-33: Smoke Concentration in ICFMP BE #3, Open-Door Tests

Table A-15: Relative Differences for Smoke Concentration in ICFMP BE #3 Tests

	ΔE (mg/m^3)	ΔM (mg/m^3)	Relative Difference
ICFMP 3-1	41.50	258.62	523%
ICFMP 3-7	55.05	250.39	355%
ICFMP 3-2	128.00	305.08	138%
ICFMP 3-8	99.53	359.64	261%
ICFMP 3-4	79.90	196.32	146%
ICFMP 3-10	70.75	194.81	175%
ICFMP 3-13	223.51	282.84	27%
ICFMP 3-16	139.07	219.10	58%
ICFMP 3-17	353.09	1453.17	312%
ICFMP 3-3	118.03	127.06	8%
ICFMP 3-9	117.00	126.03	8%
ICFMP 3-5	87.34	91.36	5%
ICFMP 3-14	91.30	126.62	39%
ICFMP 3-15	123.71	126.65	2%
ICFMP 3-18	110.18	126.08	14%

A.7 Compartment Pressure

Experimental measurements for room pressure are available from the ICFMP BE #3 test series only. The pressure within the compartment was measured at a single point, near the floor. In the simulations of the closed-door tests, the compartment is assumed to leak via a small opening near the ceiling. In order to reflect the actual leakage area in the model, the measured area was divided by 0.68, which is the orifice flow coefficient used in MAGIC for all flows through vertical openings.

A.7.1 ICFMP BE #3

Visual examination of experimental data and model results plots strongly suggest that in tests with open doors, where leakages are not critical because of the large door opening, MAGIC captures both the magnitude and the profile of the pressure. These figures describe a negative pressure profile at the floor of the room, indicating that fresh air is moving into the enclosure.

Similarly, in closed-door tests, MAGIC is able to capture the both peaks and pressure profiles. It is important to mention that fan tests were conducted before some of the tests, resulting in relatively well-known leakage areas. Furthermore, notice that MAGIC captures the positive and negative pressure peaks. These peaks are an indication of a positively pressurized room in the early stages of the test, and a negatively pressurized room when the fuel supply is discontinued and heat losses to the boundaries are higher than the heat release rate of the fire.

Comparisons between measurements and predictions are shown in Figure A-34 and Figure A-35. For tests in which the door to the compartment is open, the over-pressures are only a few Pascals at the early stages of the fire; however, when the door is closed, the over-pressures can be up to several hundred Pascals. The calculated relative differences are listed in Table A-16.

The relative differences were calculated as follows:

- For closed-door rooms, the relative difference refers to the positive peak at the early stages of the fire. Positive relative differences indicate that MAGIC over-predicted the measured peak.

- For open-door rooms, the relative differences refer to the negative magnitudes of the pressure, typically at the late stages of the test. Positive relative differences suggest that MAGIC calculated a lower pressure than the experimental measurement.

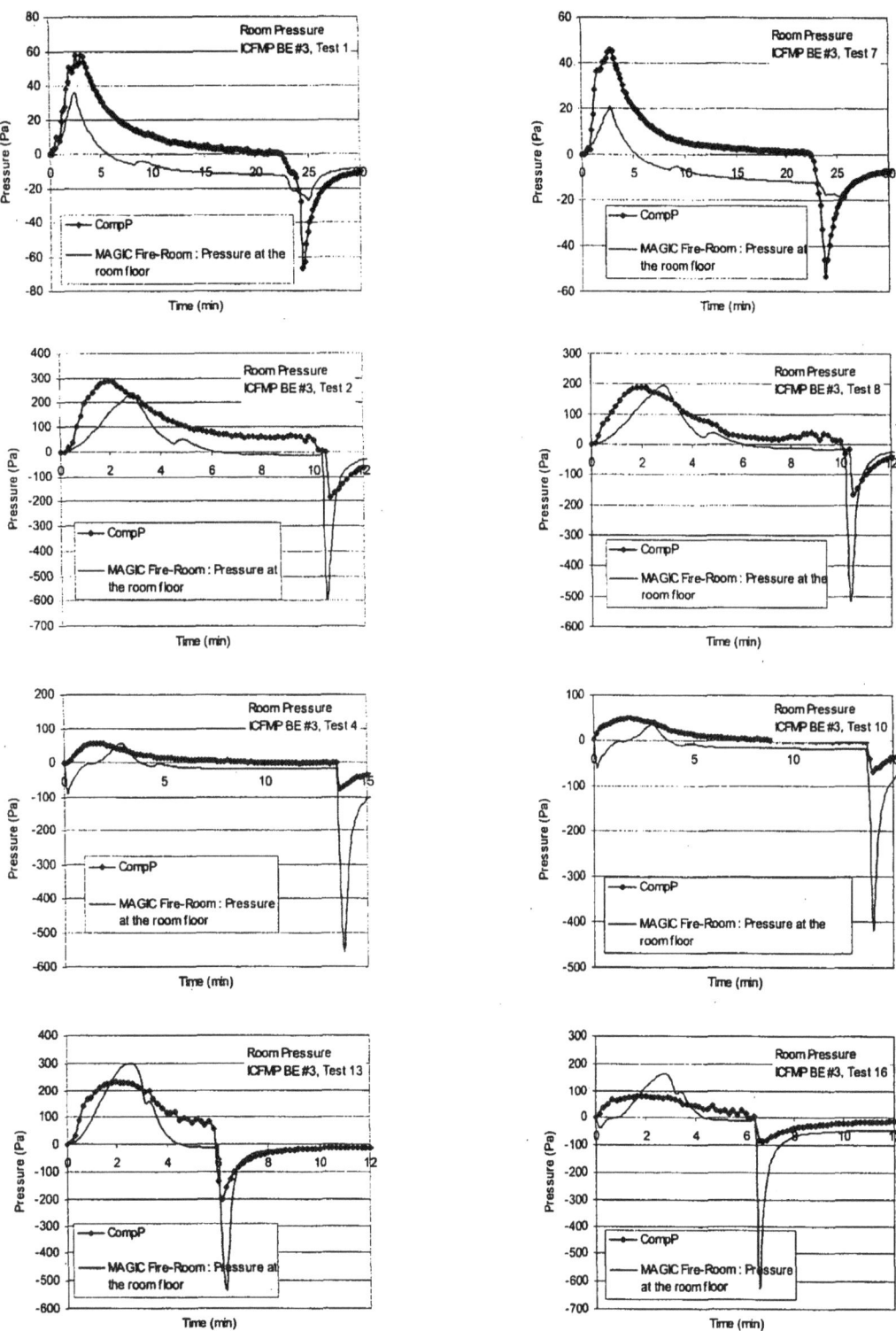

Figure A-34: Compartment Pressure in ICFMP BE #3, Closed-Door Tests

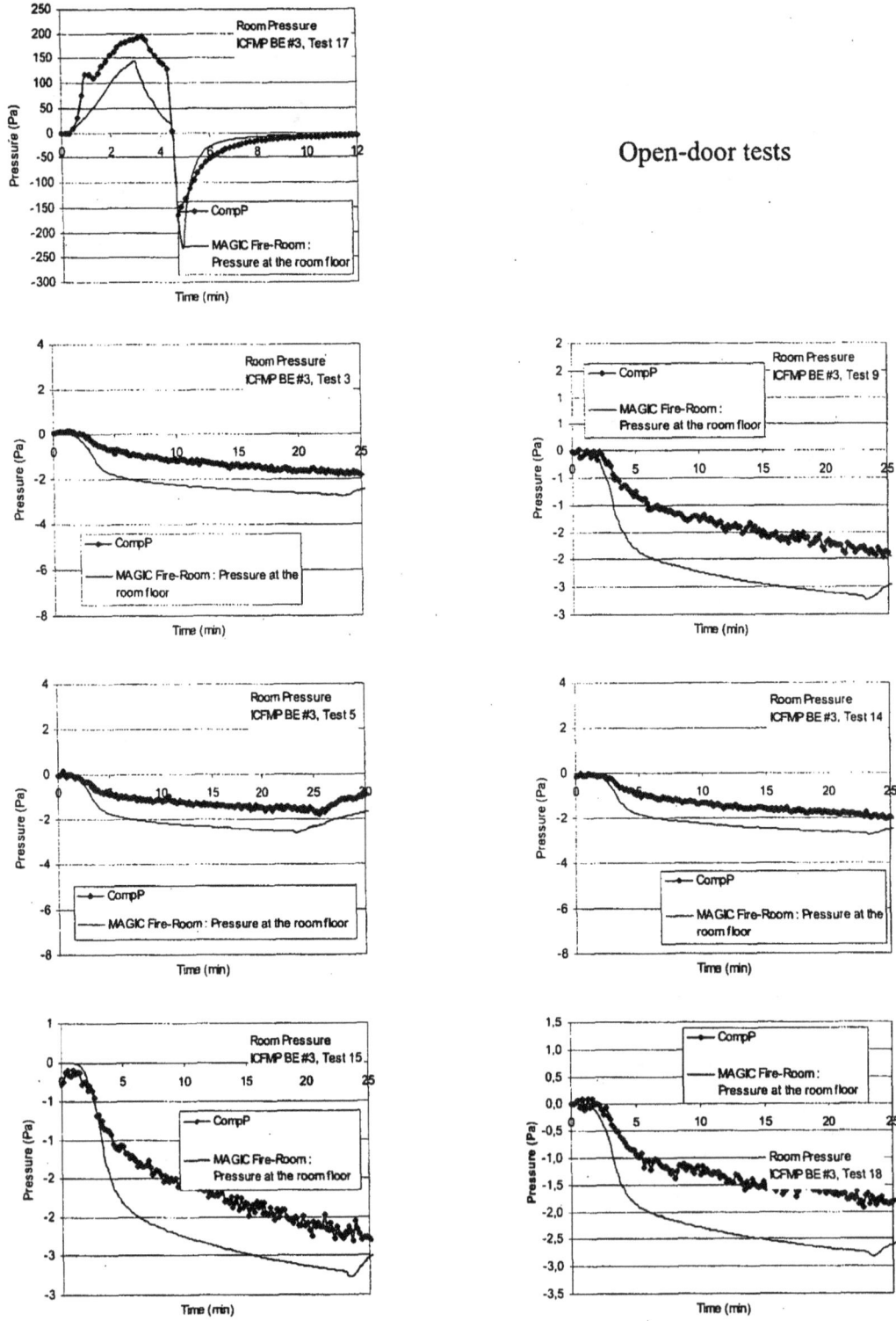

Figure A-35: Compartment Pressure in ICFMP BE #3, Open-Door Tests

Table A-16: Relative Differences for Compartment Pressure in ICFMP BE #3 Tests

	ΔE (Pa)	ΔM (Pa)	Relative Difference
ICFMP 3-1	57.6	35.7	-38%
ICFMP 3-7	45.9	21.0	-54%
ICFMP 3-2	290.0	241.2	-17%
ICFMP 3-8	189.3	195.7	3%
ICFMP 3-4	56.6	57.6	2%
ICFMP 3-10	49.3	34.3	-30%
ICFMP 3-13	231.5	313.6	35%
ICFMP 3-16	80.6	162.5	102%
ICFMP 3-17	194.9	144.8	-26%
ICFMP 3-3	-1.9	-2.7	41%
ICFMP 3-9	-2.0	-2.7	36%
ICFMP 3-5	-1.8	-2.5	40%
ICFMP 3-14	-2.1	-2.7	31%
ICFMP 3-15	-2.4	-2.8	17%
ICFMP 3-18	-2.0	-2.8	41%

A.8 Target Temperature and Heat Flux

Target temperature and heat flux data are available from ICFMP BE #3, #4, and #5. In BE #3, the targets are various types of cables in various configurations — horizontal, vertical, in trays, or free-hanging. In BE #4, the targets are three rectangular slabs of different materials instrumented with heat flux gauges and thermocouples. In BE #5, the targets are again cables, in this case, bundled power and control cables in a vertical ladder.

Cable targets in MAGIC can be represented as cables or thermal targets. This section provides graphical comparisons and relative differences for both options. Targets are virtual sensors in the computational domain, characterized by a thickness and themo-physical properties. By contrast, cables are specified as cylinders of some length, with multiple concentric layers of different materials to account for jacket, insulation, and conductor.

For radiated heat flux, the Target/Heat Flux/Incident Heat Flux output option in MAGIC was selected for comparison with experimental results. This is the sum of all external radiative heat fluxes impacting the target. In the case of total heat flux, the Target/Heat Flux/Total Heat Flux "Flux Meter" output option in MAGIC was selected for comparison.

The total flux gauges used in the experiment correspond to the total heat flux with a target calibrated using cooling water with a temperature of 75 °C (167 °F). In MAGIC, the target measuring total heat flux is based on the ambient temperature of ~20 °C (68 °F).

A.8.1 ICFMP BE #3

For each of the four cable targets considered, measurements of the local gas temperature, surface temperature, radiative heat flux, and total heat flux are available. The following pages display comparisons of these quantities for Control Cable B, Horizontal Cable Tray D, Power Cable F, and Vertical Cable Tray G.

The superposition of gas temperature, heat flux, and surface temperature in the figures on the following pages provides information about how cables heat up in fires. In MAGIC, cables can be modeled using the Target option or the Source/Cable option. This study evaluates both. The Target option is listed in the graphical comparisons as "Surface Temperature of the Target." The Cable/Source option is listed as "Maximum Surface Temperature."

Favorable or unfavorable predictions of cable surface temperatures can usually be explained in terms of comparable errors in the prediction of the thermal environment in the vicinity of the cable. Regardless of the complexity of the target, the model must be able to predict the thermal insult to it.

The following figures and tables provide the graphical comparisons and calculated relative differences. Results are classified by cable. That is, for a selected cable, the surface and gas temperature and the total and radiated heat fluxes for the 15 tests are grouped together. The tables for relative differences for those 15 tests follow each group of graphical comparisons. Relative differences were calculated for surface temperature, radiative heat flux, and total heat flux.

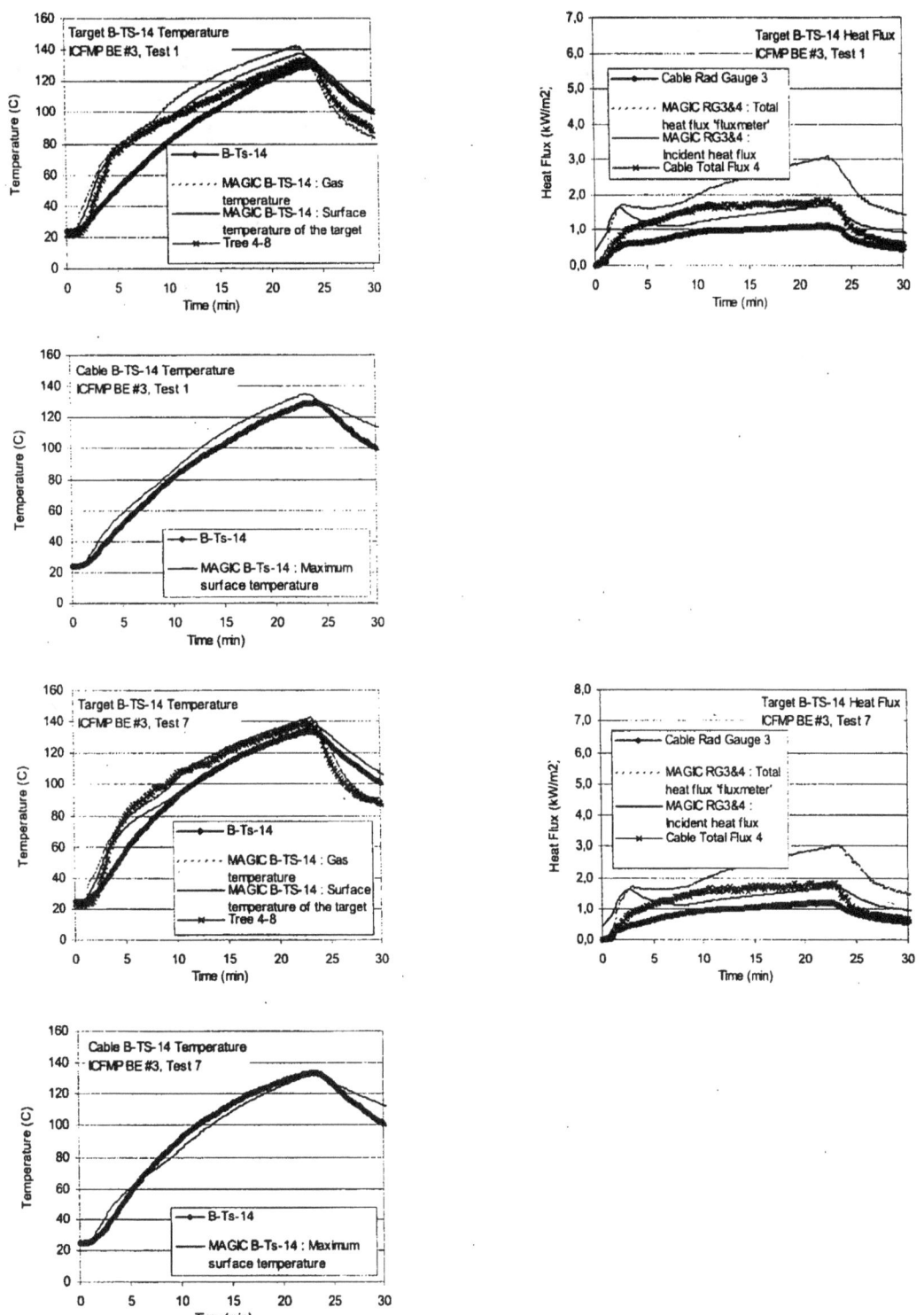

Figure A-36: Thermal Environment near Cable B, ICFMP BE #3, Tests 1 and 7

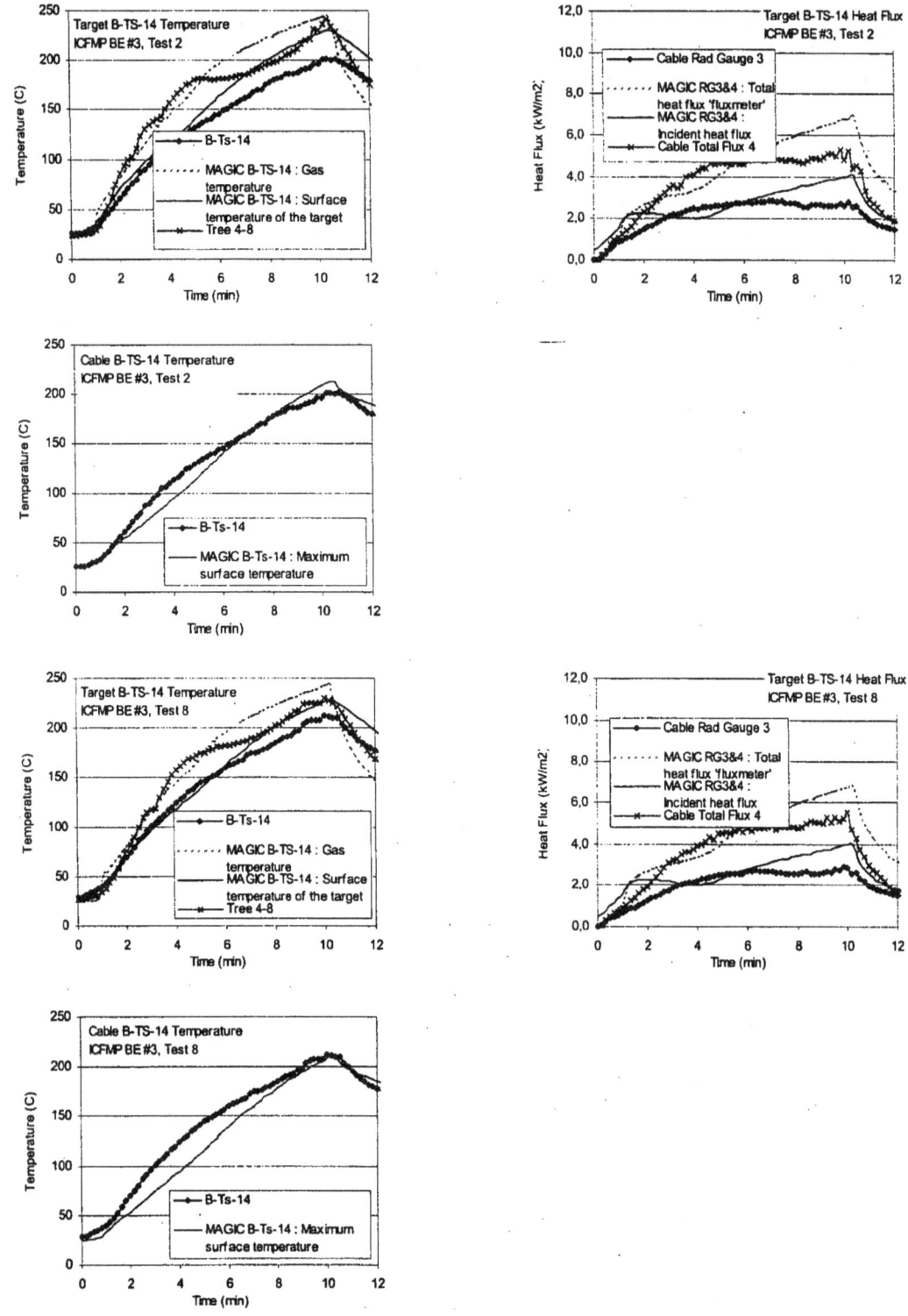

Figure A-37: Thermal Environment near Cable B, ICFMP BE #3, Tests 2 and 8

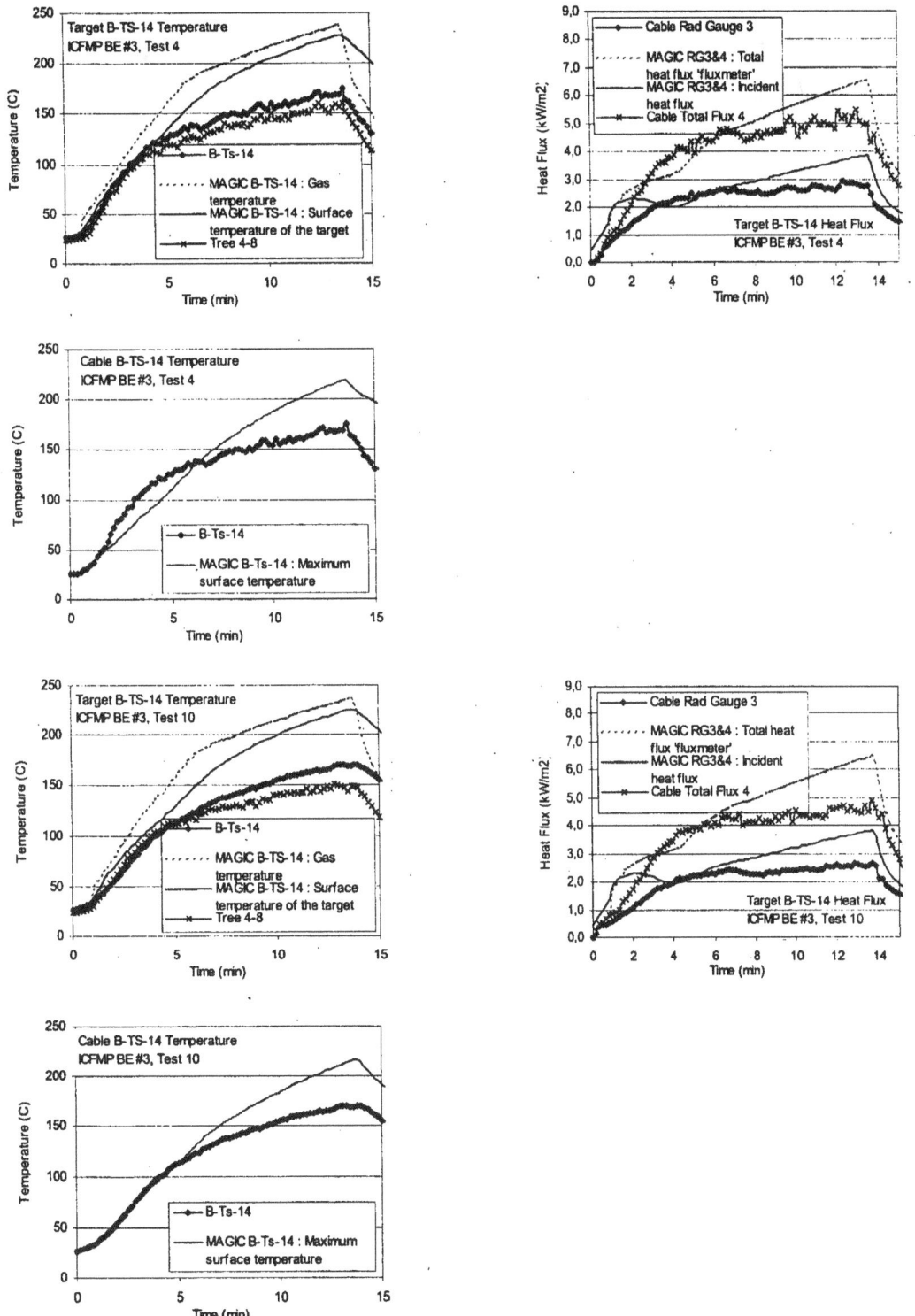

Figure A-38: Thermal Environment near Cable B, ICFMP BE #3, Tests 4 and 10

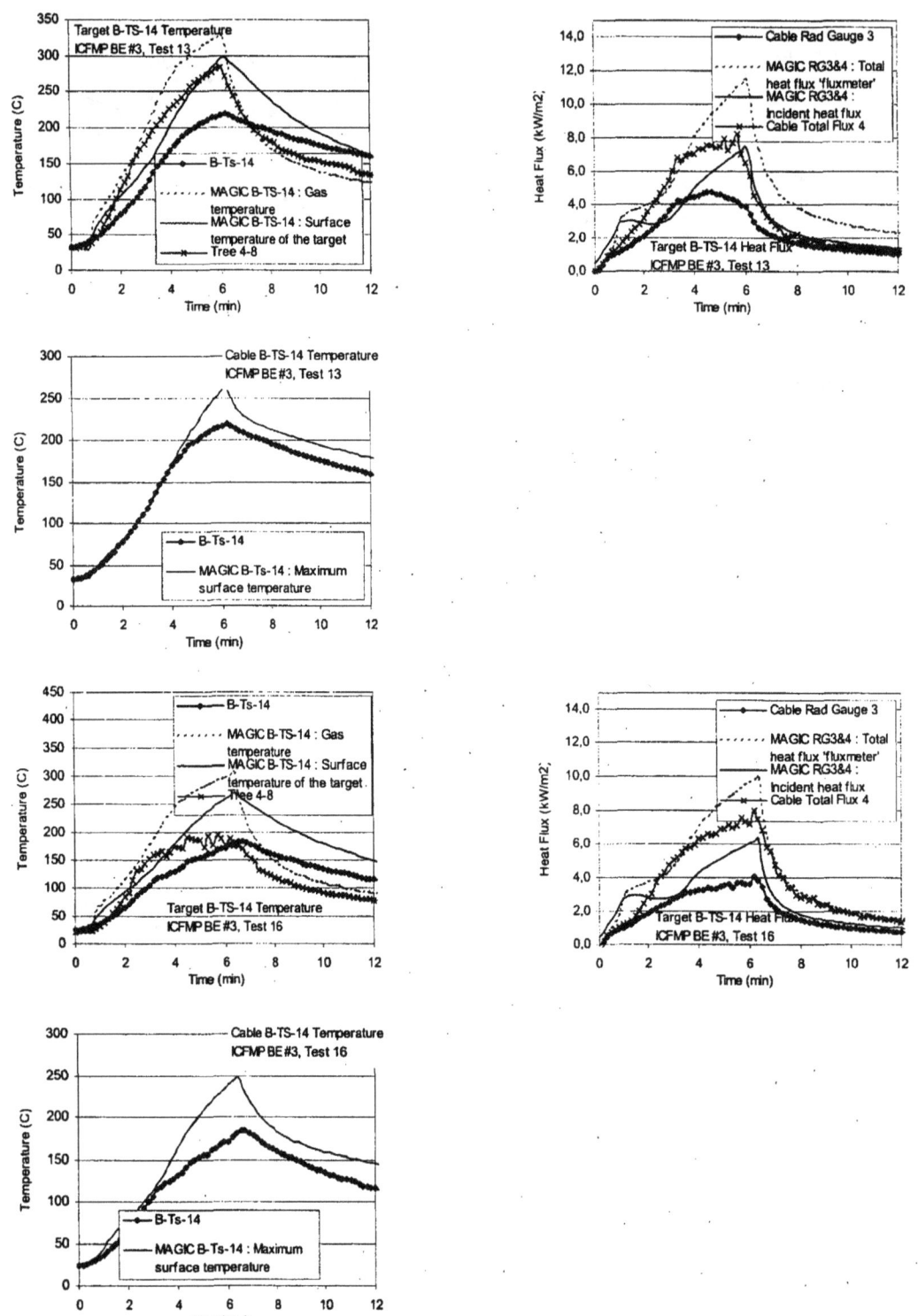

Figure A-39: Thermal Environment near Cable B, ICFMP BE #3, Tests 13 and 16

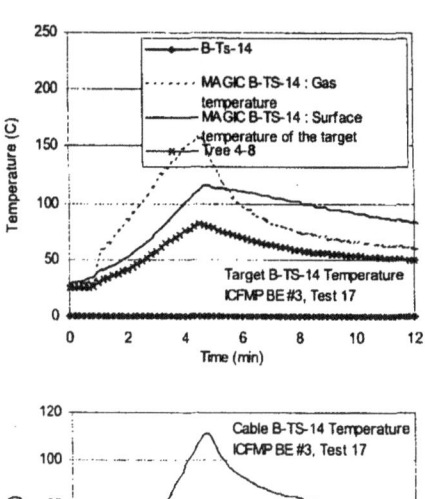

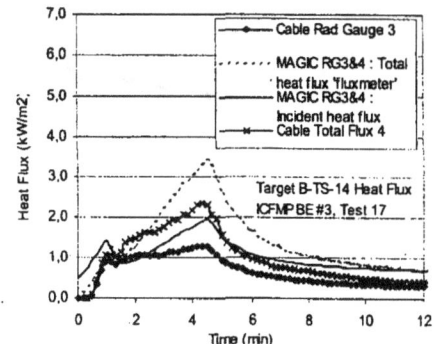

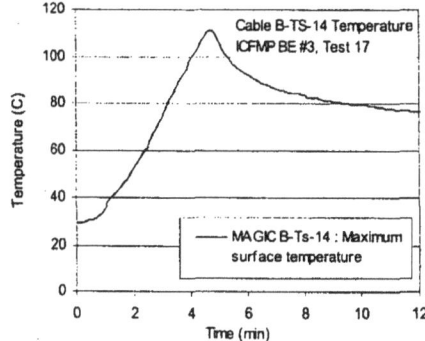

Figure A-40: Thermal Environment near Cable B, ICFMP BE #3, Test 17

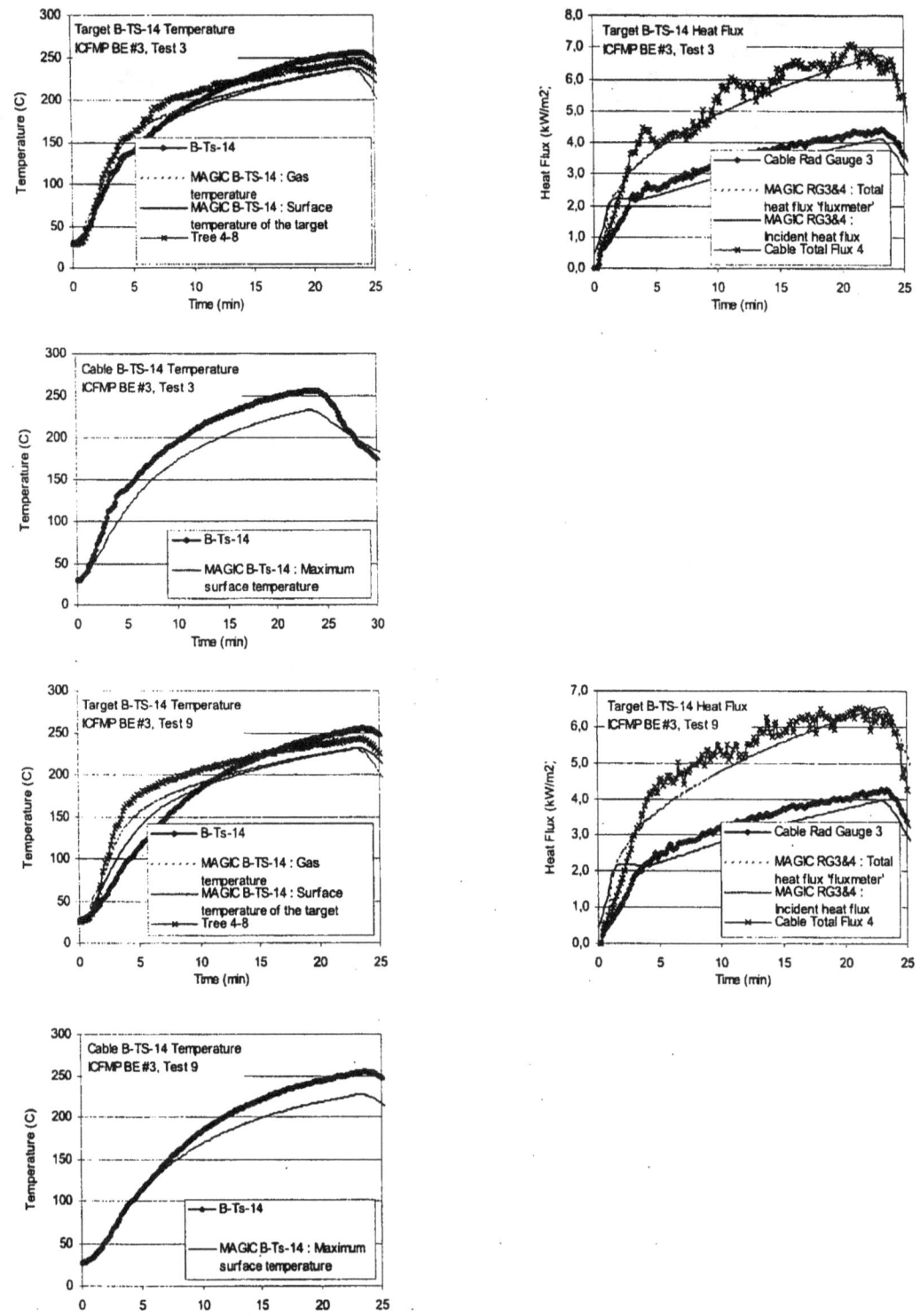

Figure A-41: Thermal Environment near Cable B, ICFMP BE #3, Tests 3 and 9

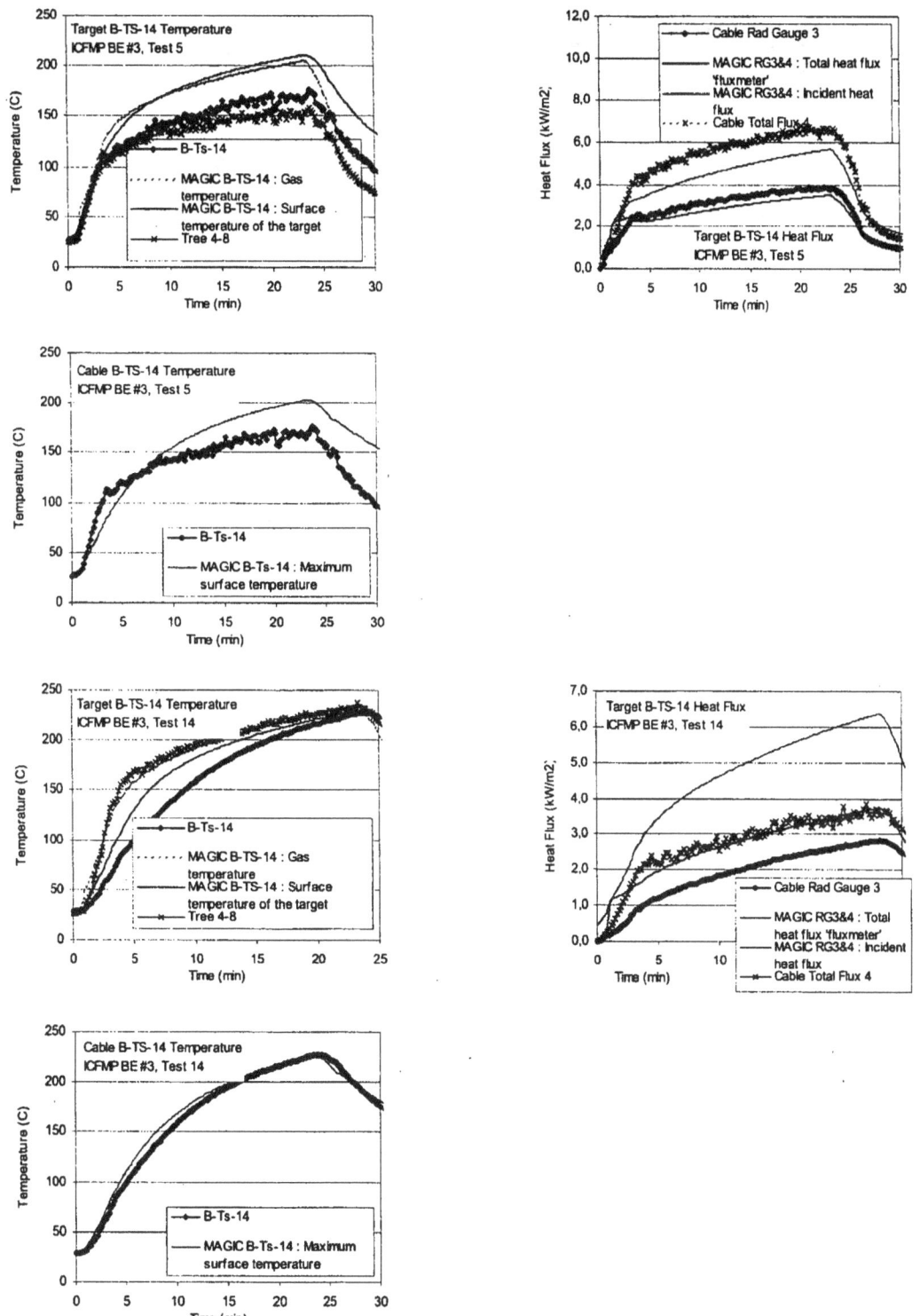

Figure A-42: Thermal Environment near Cable B, ICFMP BE #3, Tests 5 and 14

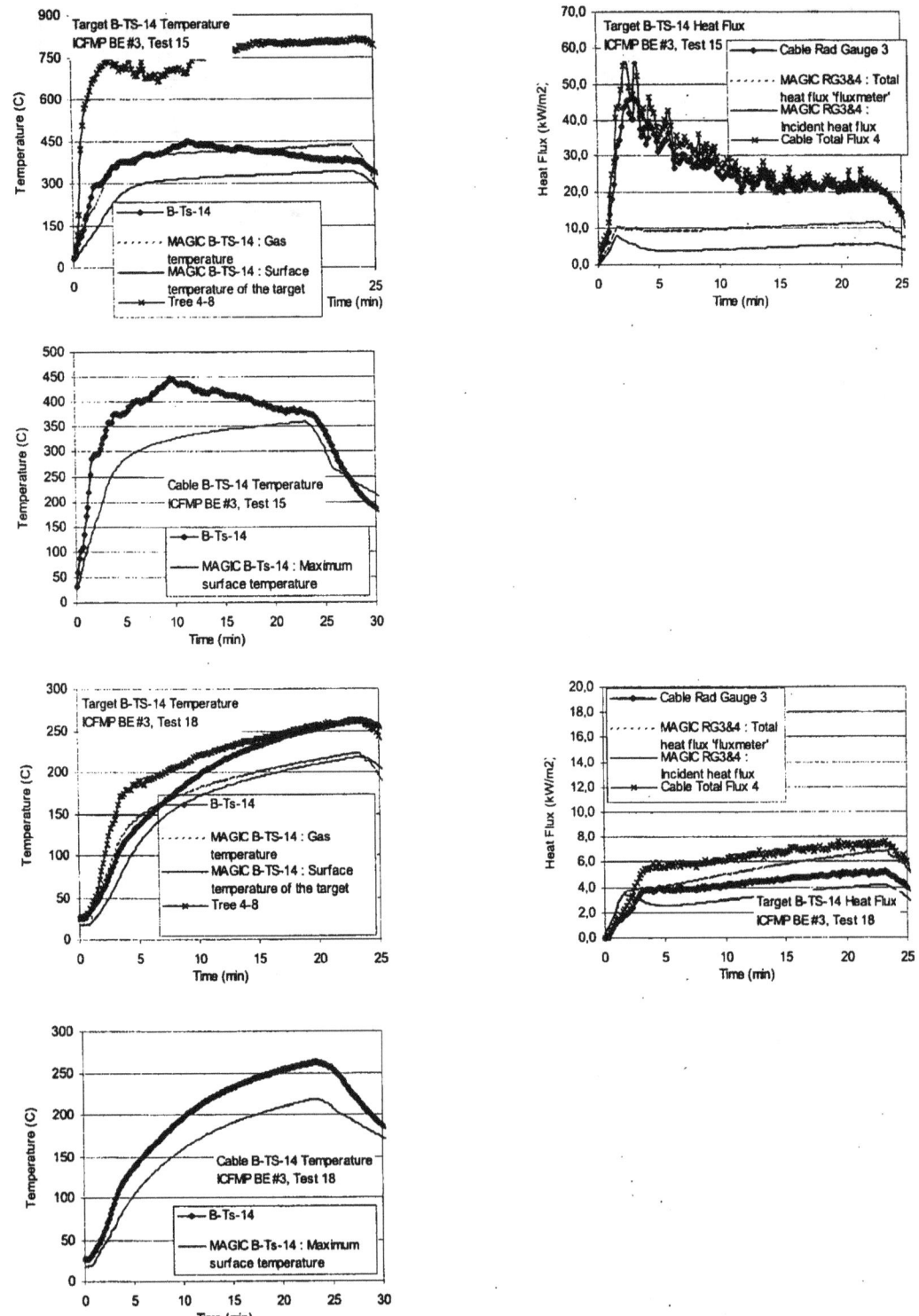

Figure A-43: Thermal Environment near Cable B, ICFMP BE #3, Tests 15 and 18

Table A-17: Relative Differences for Surface Temperature of Cable B

Control Cable	Target Surface Temp, B-TS-14			Cable Surface Temp, B-TS-14		
	ΔE (°C)	ΔM (°C)	Relative Diff	ΔE (°C)	ΔM (°C)	Relative Diff
Test 1	106	115	8%	106	112	5%
Test 7	90	111	23%	87	108	-1%
Test 2	176	204	16%	176	187	6%
Test 8	183	203	11%	183	186	1%
Test 4	149	201	35%	149	192	29%
Test 10	144	198	38%	144	189	31%
Test 13	186	264	42%	186	228	25%
Test 16	160	244	52%	160	222	38%
Test 17		83			80	
Test 3	226	206	-9%	226	202	-11%
Test 9	228	204	-10%	228	199	-12%
Test 5	150	182	21%	150	175	16%
Test 14	199	201	1%	199	199	0%
Test 15	416	315	-24%	416	333	-20%
Test 18	236	200	-15%	236	200	-15%

Table A-18: Relative Differences for Radiative and Total Heat Flux to Cable B

Control Cable	Radiant Heat Flux Gauge 3			Total Heat Flux, Gauge 4		
	ΔE (kW/m^2)	ΔM (kW/m^2)	Relative Diff	ΔE (kW/m^2)	ΔM (kW/m^2)	Relative Diff
Test 1	1.1	1.3	12%	1.85	3.06	65%
Test 7	2.9	3.6	26%	1.84	2.99	62%
Test 2	4.4	3.6	-18%	5.26	6.92	31%
Test 8	2.9	3.4	17%	5.58	6.85	23%
Test 4	3.9	3.0	-22%	5.52	6.57	19%
Test 10	1.2	1.2	2%	4.91	6.51	33%
Test 13	2.9	3.6	23%	8.26	11.48	39%
Test 16	4.3	3.5	-19%	8.37	10.03	20%
Test 17	2.7	3.4	25%	2.36	3.43	45%
Test 3	4.8	7.0	47%	7.10	6.74	-5%
Test 9	2.8	3.3	15%	6.58	6.56	0%
Test 5	46.5	7.4	-84%	6.86	5.67	-17%
Test 14	4.1	5.9	42%	3.82	6.37	65%
Test 15	1.3	1.5	15%	57.72	11.72	-80%
Test 18	5.2	3.8	-28%	7.61	6.86	-10%

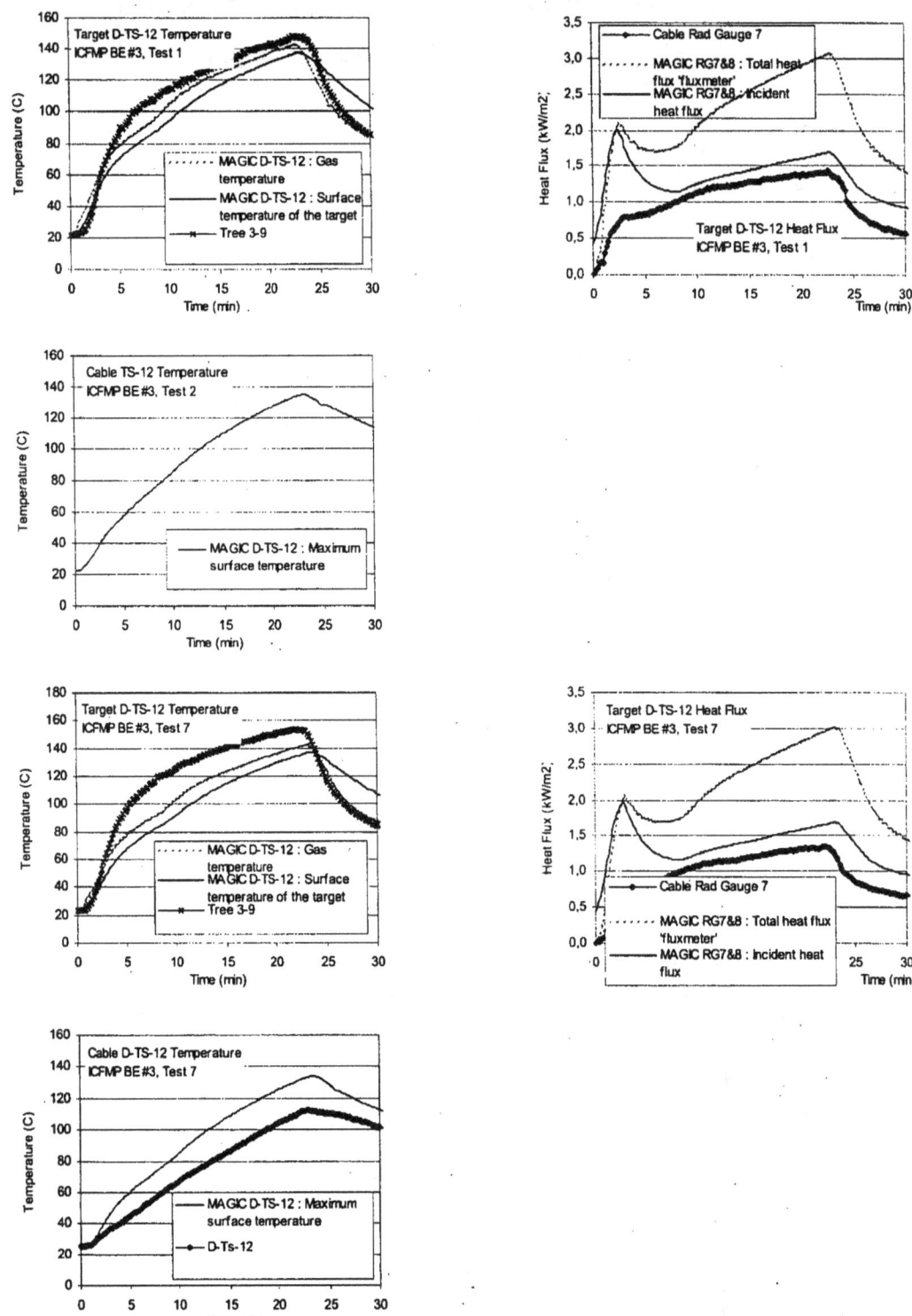

Figure A-44: Thermal Environment near Cable D, ICFMP BE #3, Tests 1 and 7

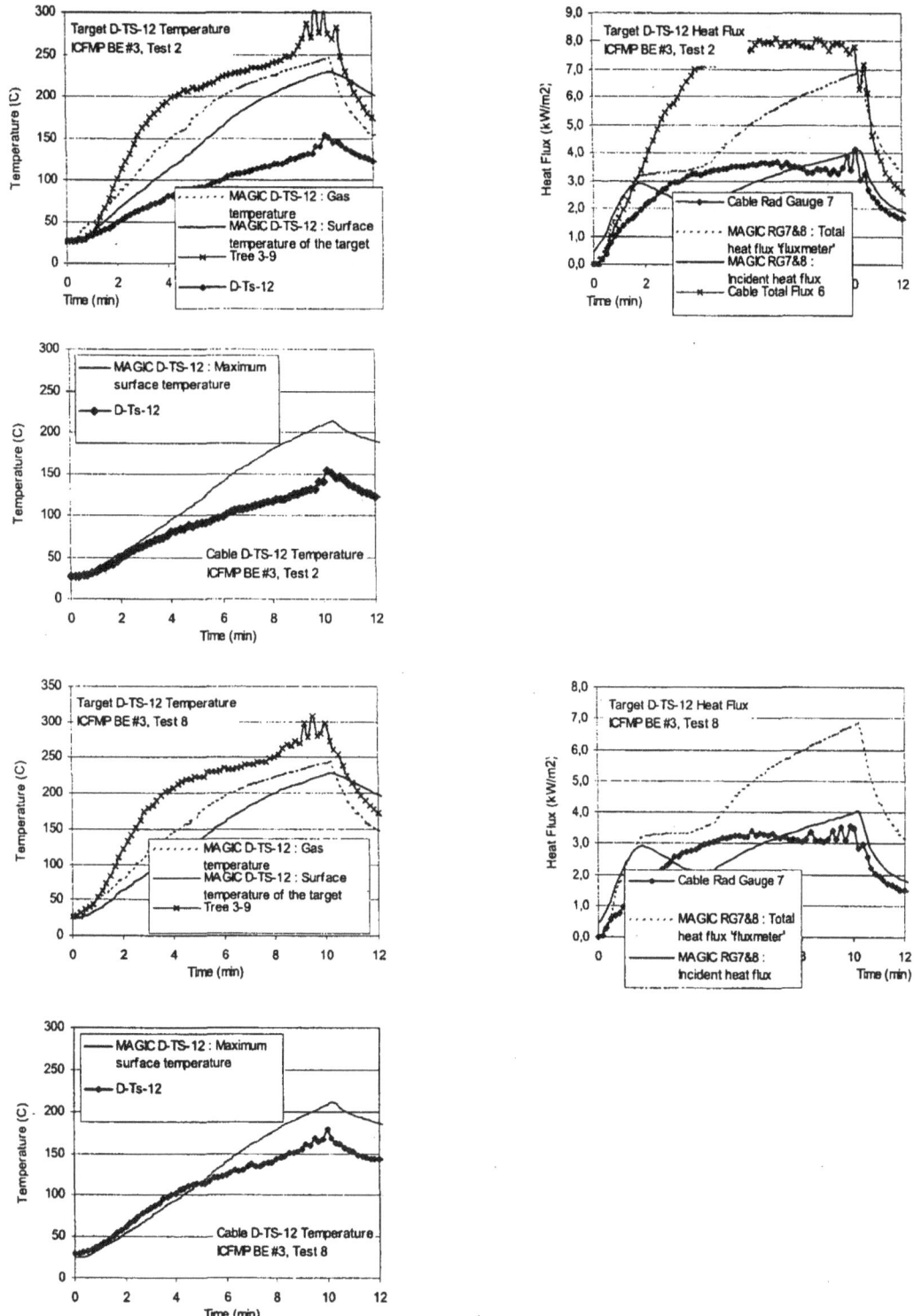

Figure A-45: Thermal Environment near Cable D, ICFMP BE #3, Tests 2 and 8

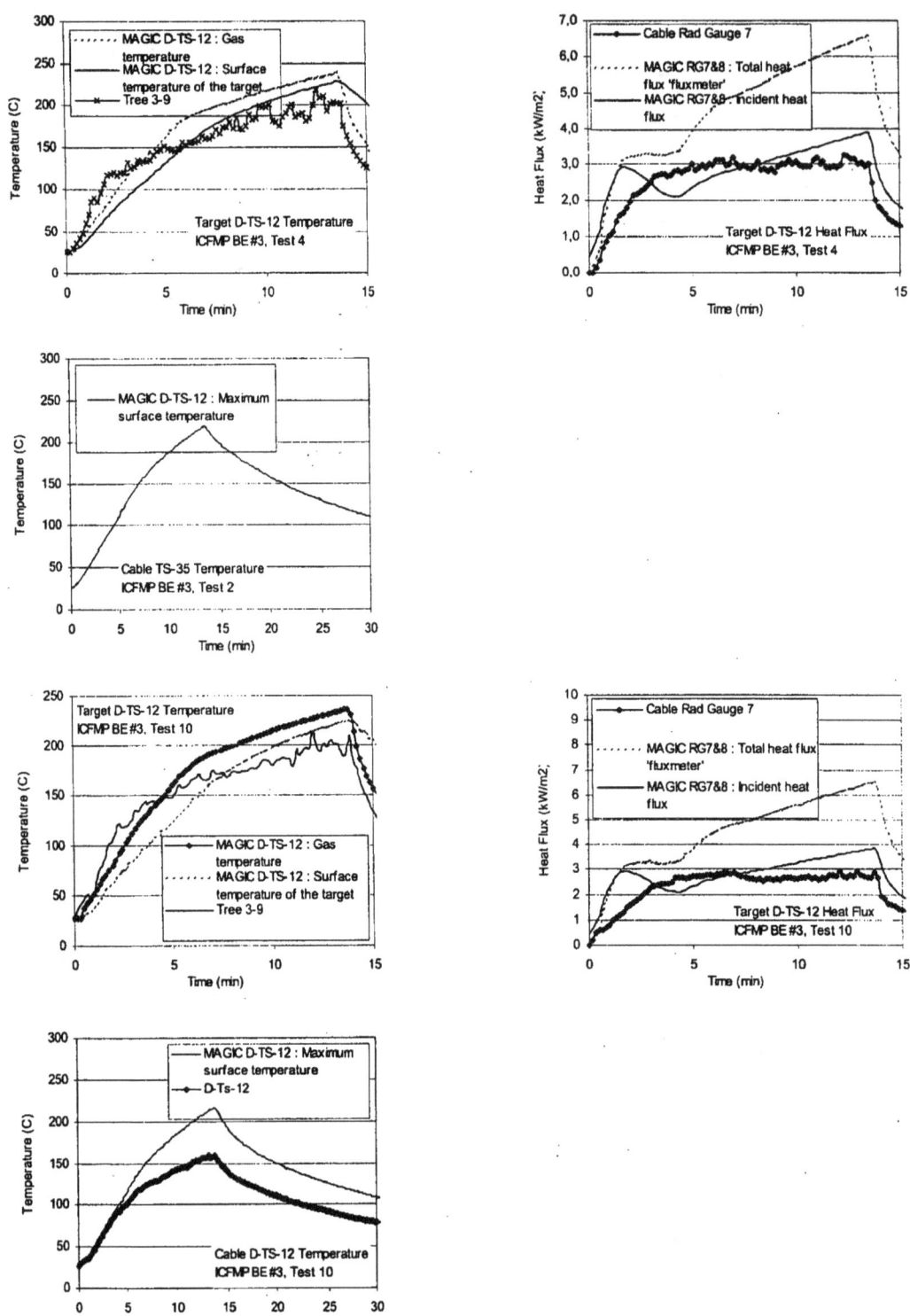

Figure A-46: Thermal Environment near Cable D, ICFMP BE #3, Tests 4 and 10

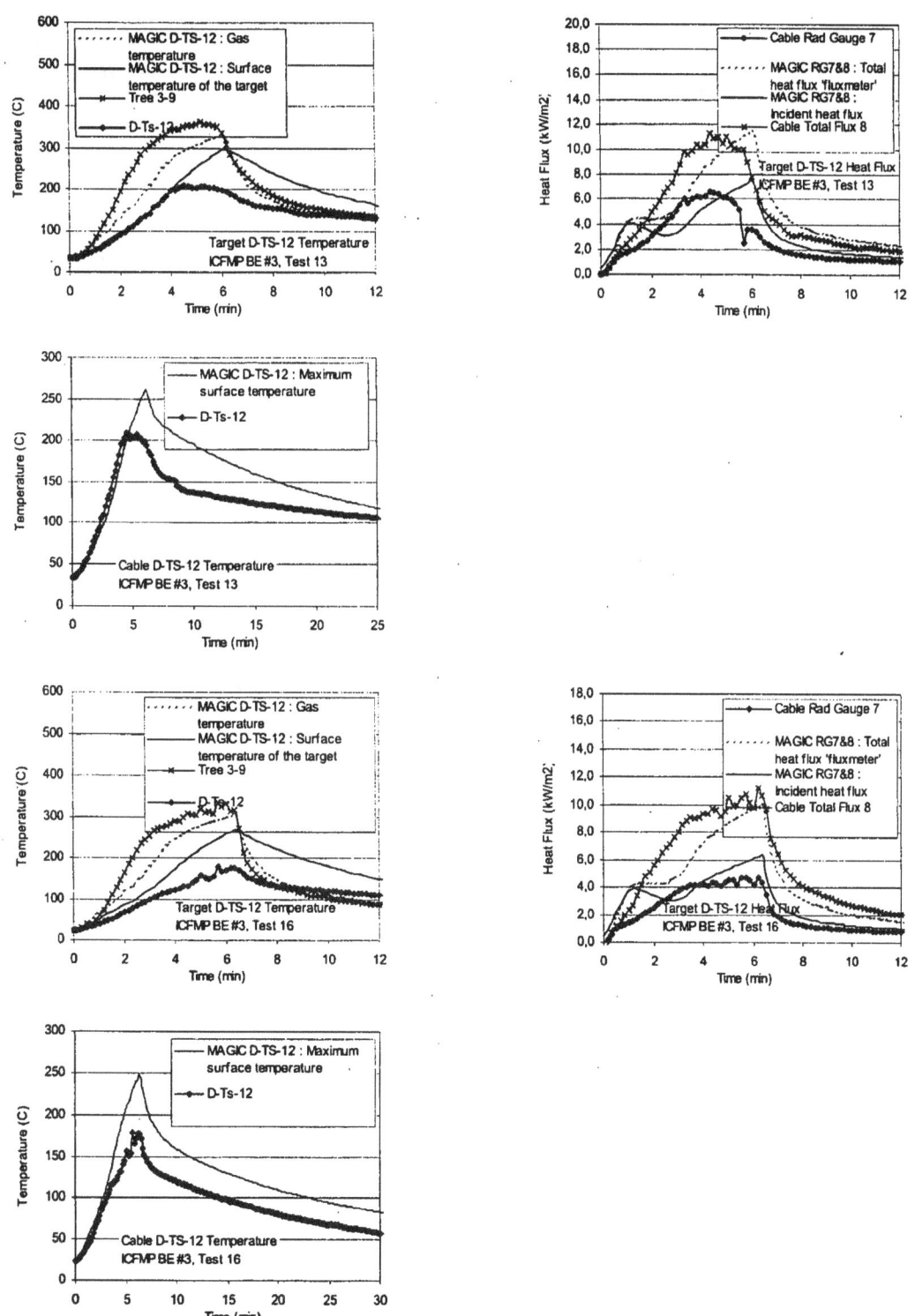

Figure A-47: Thermal Environment near Cable D, ICFMP BE #3, Tests 13 and 16

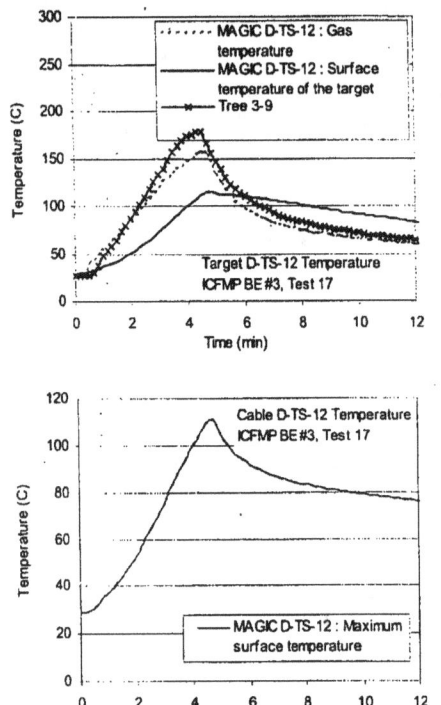 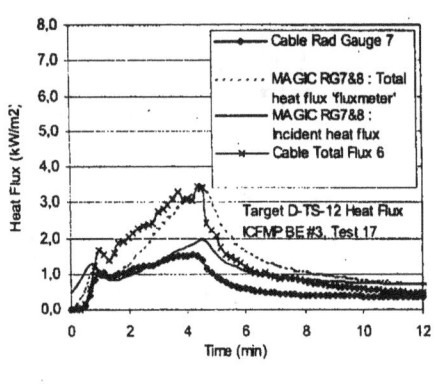

Figure A-48: Thermal Environment near Cable D, ICFMP BE #3, Test 17

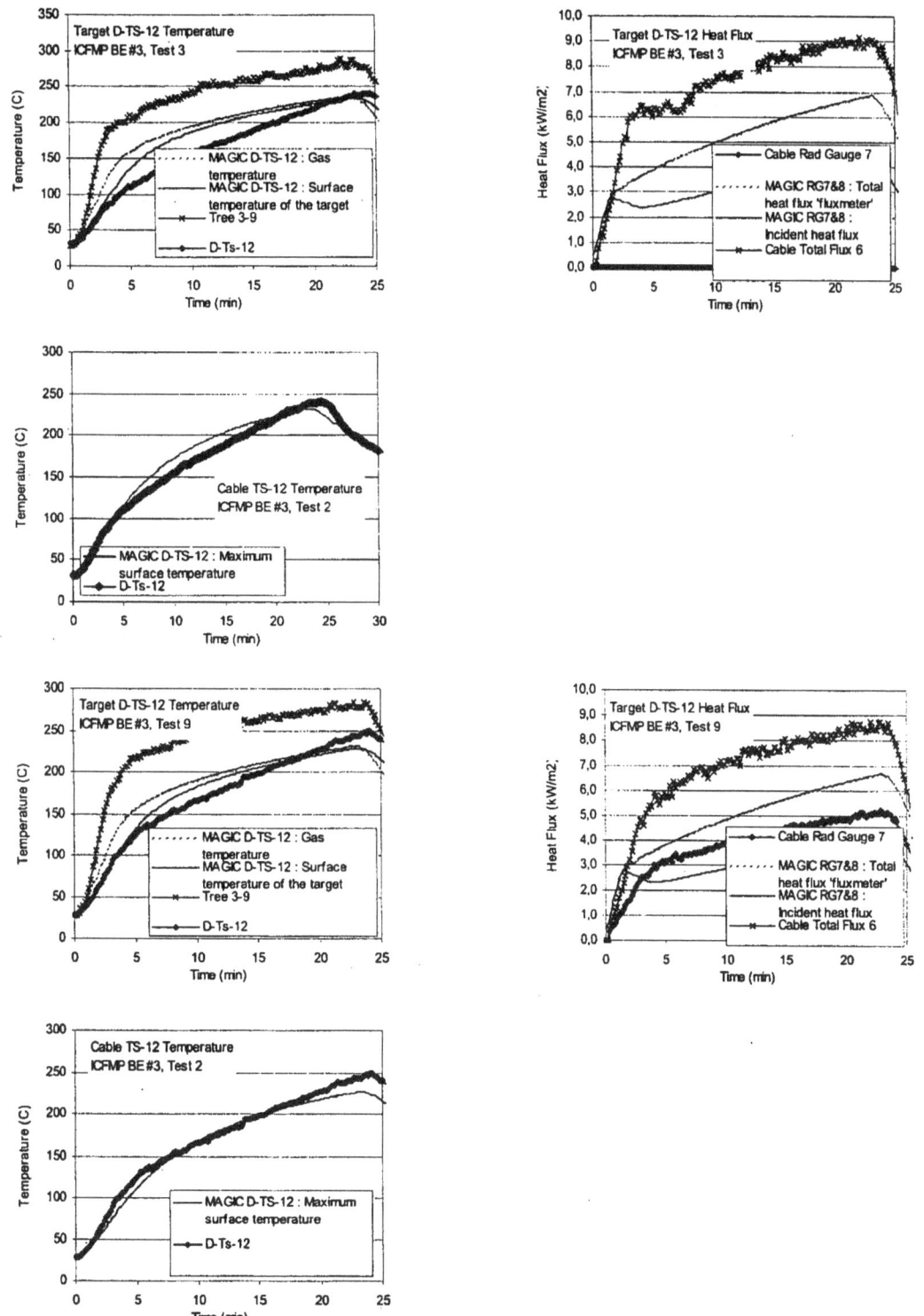

Figure A-49: Thermal Environment near Cable D, ICFMP BE #3, Tests 3 and 9

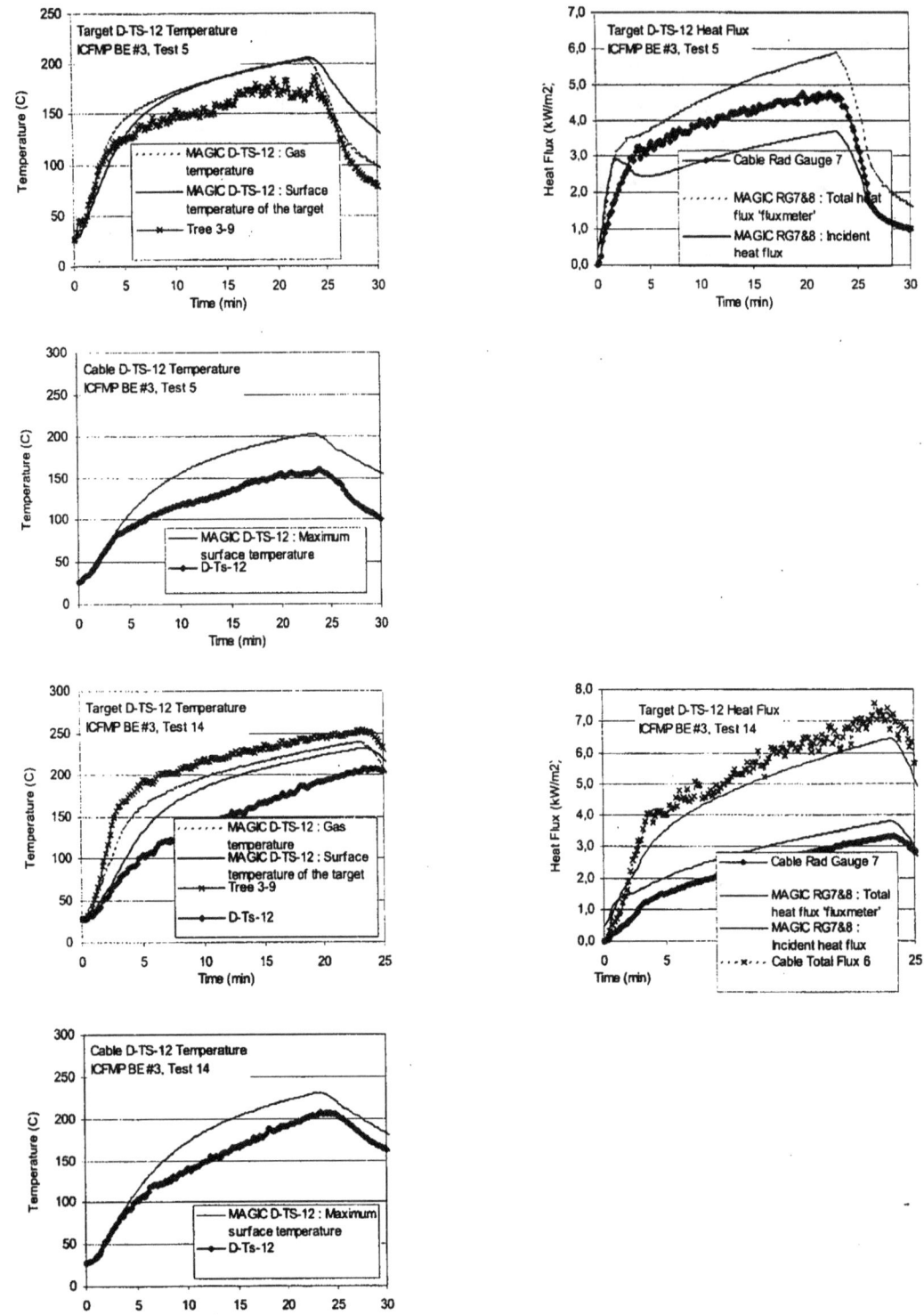

Figure A-50: Thermal Environment near Cable D, ICFMP BE #3, Tests 5 and 14

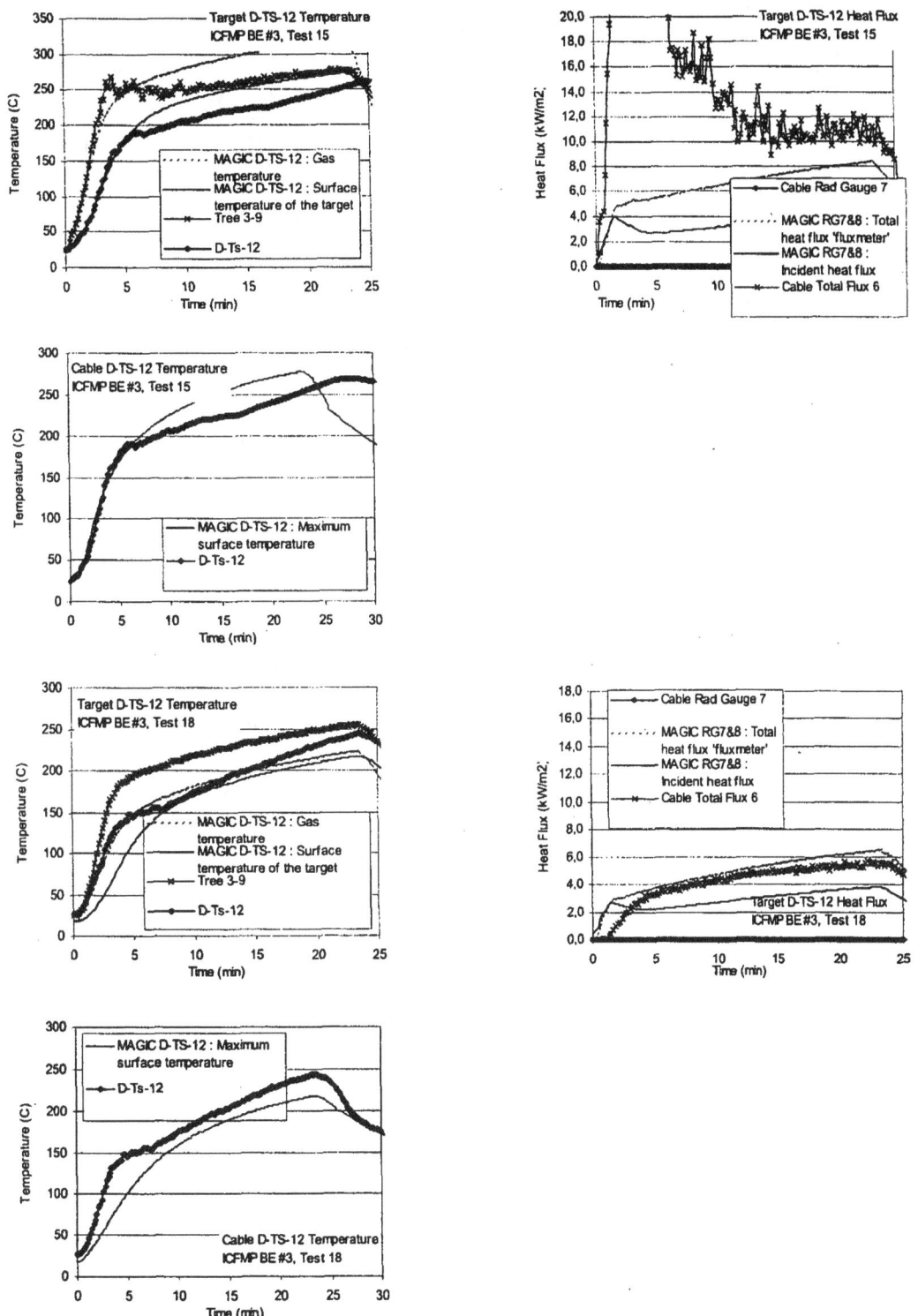

Figure A-51: Thermal Environment near Cable D, ICFMP BE #3, Tests 15 and 18

Table A-19: Relative Differences for Surface Temperature of Cable D

Control Cable	Target Surface Temp, D-TS-12			Cable Surface Temp, D-TS-12		
	ΔE (°C)	ΔM (°C)	Relative Diff	ΔE (°C)	ΔM (°C)	Relative Diff
Test 1		115			112	
Test 7	87	111	27%	87	108	-4%
Test 2	126	204	62%	126	187	48%
Test 8	150	203	36%	150	186	4%
Test 4	113	201	77%	113	192	70%
Test 10	132	198	50%	132	189	43%
Test 13	173	263	52%	173	229	32%
Test 16	156	243	56%	156	222	42%
Test 17						
Test 3	210	204	-3%	210	201	-4%
Test 9	220	201	-9%	220	199	-8%
Test 5	132	178	34%	132	174	9%
Test 14	178	203	14%	178	202	14%
Test 15	243	247	1%	243	251	8%
Test 18	217	199	-8%	217	199	-8%

Table A-20: Relative Differences for Radiative and Total Heat Flux to Cable D

Control Cable	Radiant Heat Flux Gauge 7			Total Heat Flux, Gauge 8		
	ΔE (kW/m^2)	ΔM (kW/m^2)	Relative Diff	ΔE (kW/m^2)	ΔM (kW/m^2)	Relative Diff
Test 1	1.44	1.59	10%		3.07	
Test 7	4.16	3.64	-12%	2.52	2.99	19%
Test 2		3.75		9.83	6.93	-29%
Test 8	3.26	3.43	5%	8.51	6.86	-19%
Test 4	4.78	3.22	-33%	7.23	6.60	-9%
Test 10	1.35	1.54	14%	6.71	6.54	-3%
Test 13	3.55	3.59	1%	11.22	11.50	3%
Test 16	5.26	3.61	-31%	11.67	10.08	-14%
Test 17	2.91	3.39	16%	3.29	3.43	4%
Test 3	6.58	7.02	7%	9.45	6.85	-28%
Test 9	3.32	3.33	0%	9.06	6.68	-26%
Test 14		3.95		6.07	6.45	6%
Test 5	4.83	5.90	22%	8.52	5.87	-31%
Test 15	1.52	1.49	-2%	20.87	8.39	-60%
Test 18		3.43		7.83	6.50	-17%

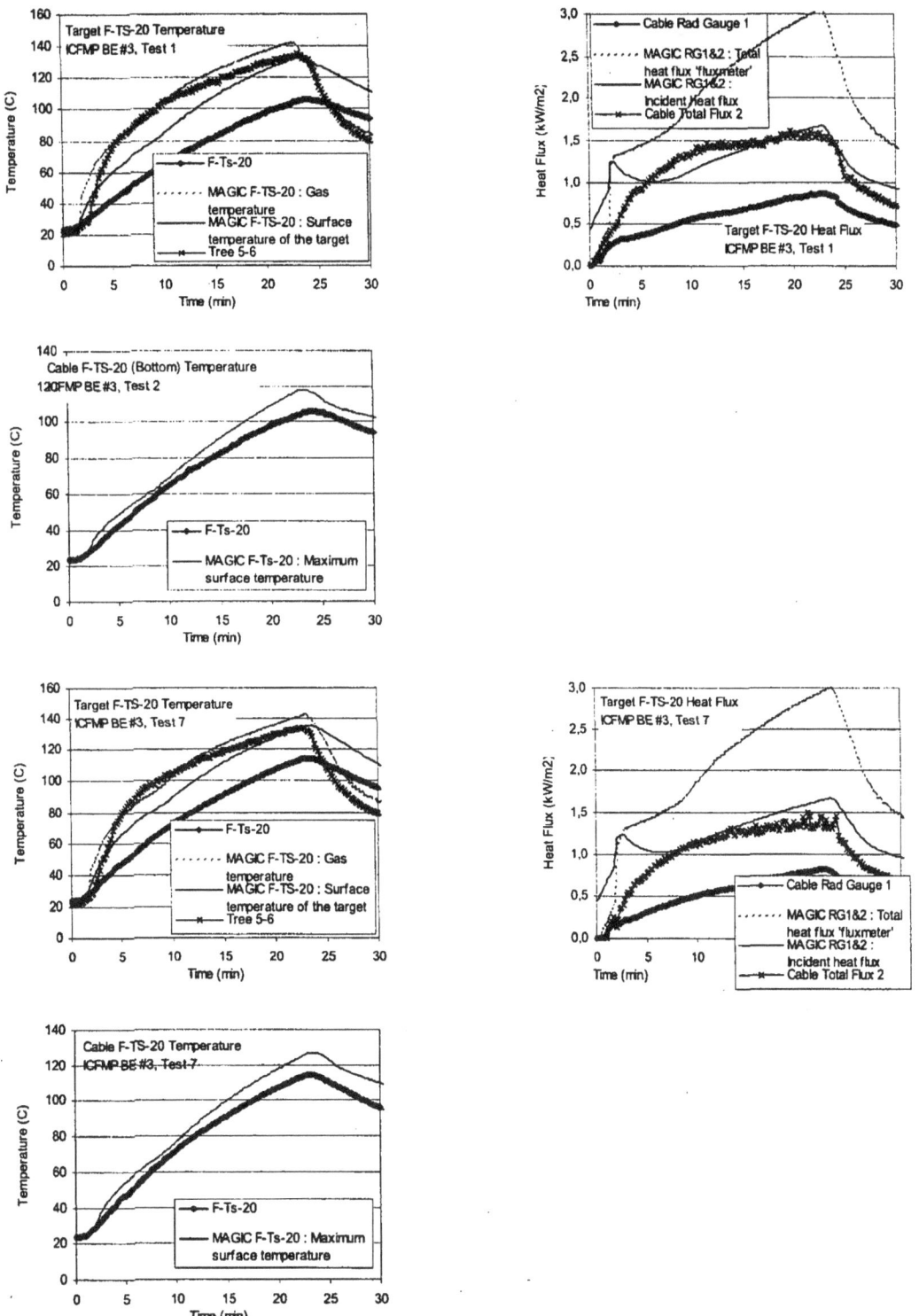

Figure A-52: Thermal Environment near Cable F, ICFMP BE #3, Tests 1 and 7

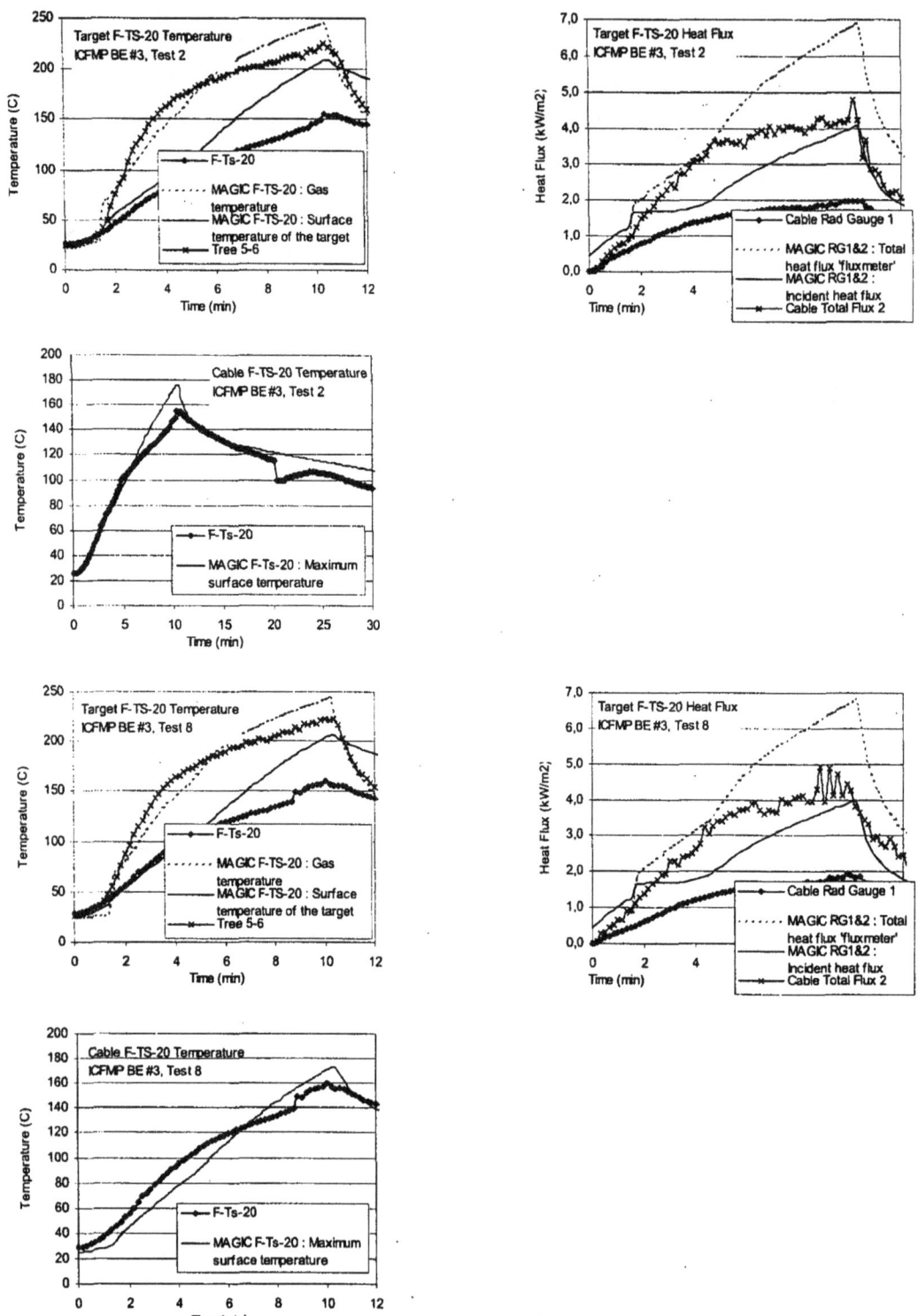

Figure A-53: Thermal Environment near Cable F, ICFMP BE #3, Tests 2 and 8

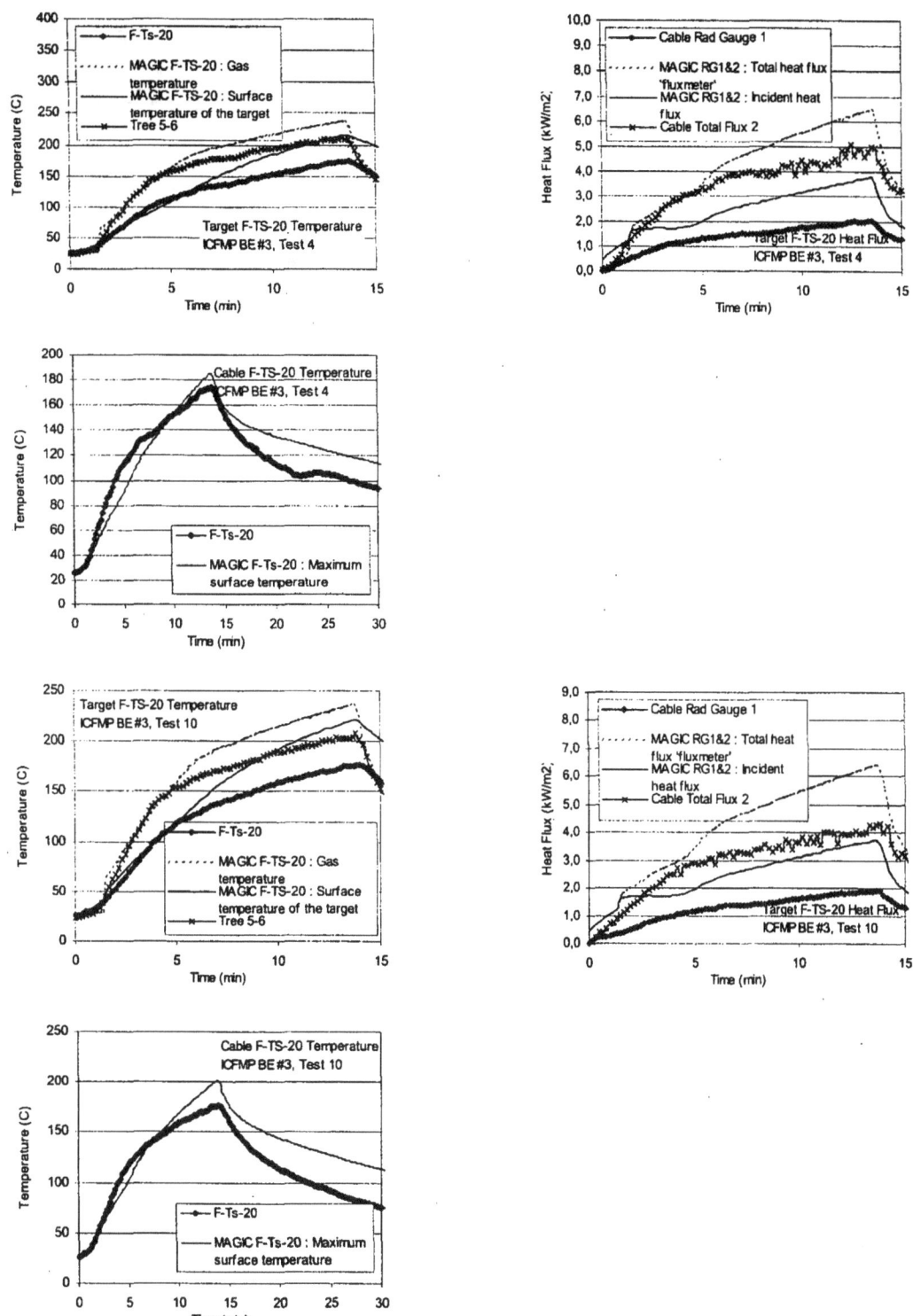

Figure A-54: Thermal Environment near Cable F, ICFMP BE #3, Tests 4 and 10

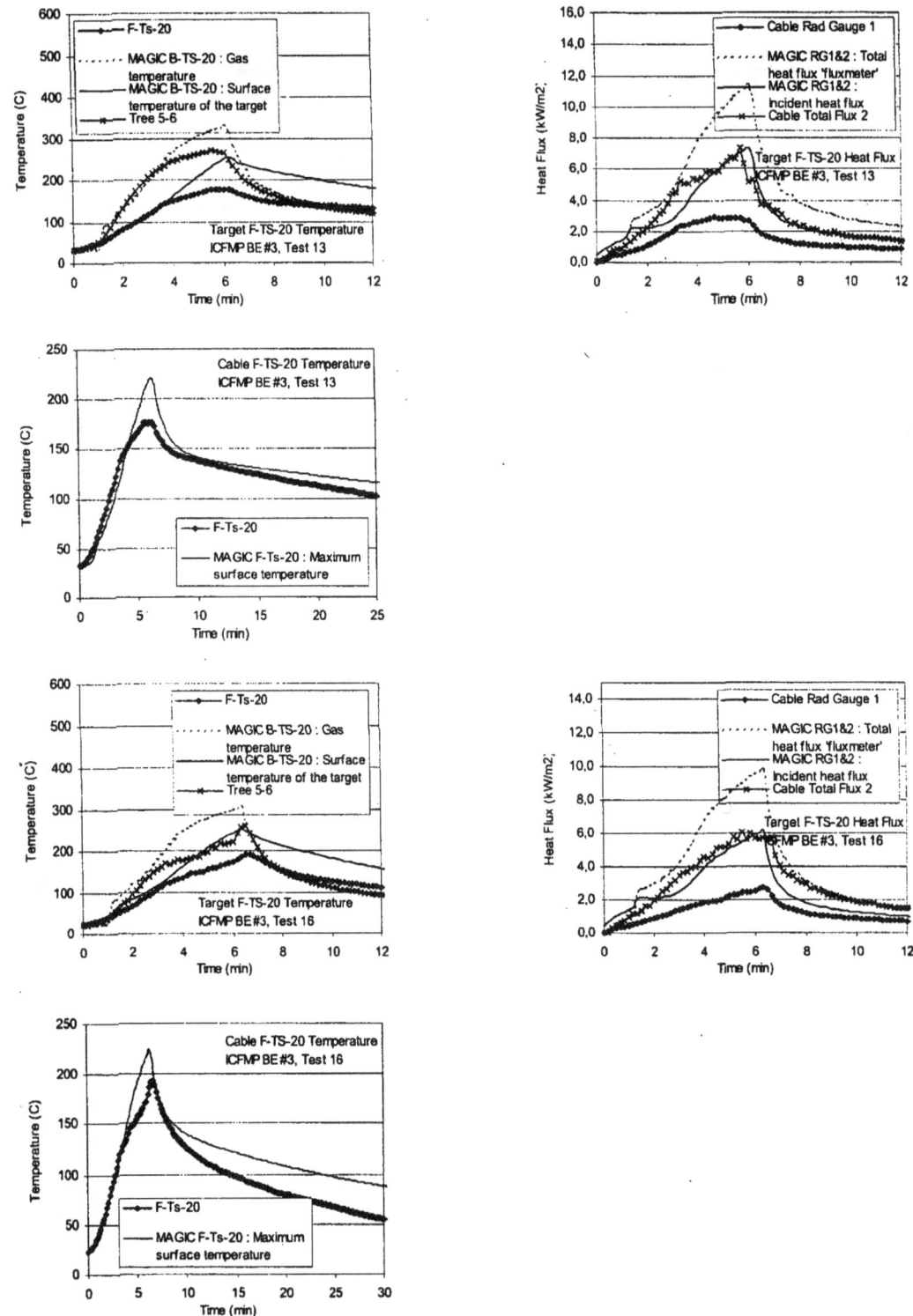

Figure A-55: Thermal Environment near Cable F, ICFMP BE #3, Tests 13 and 16

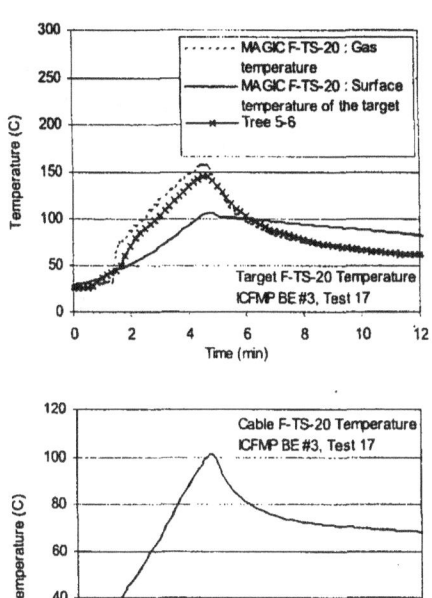

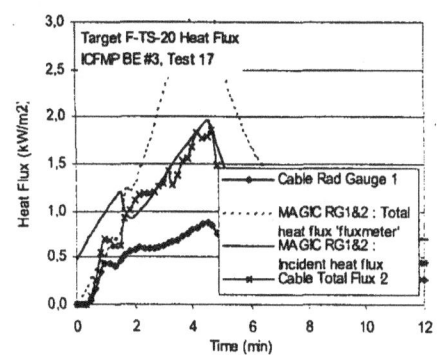

Figure A-56: Thermal Environment near Cable F, ICFMP BE #3, Test 17

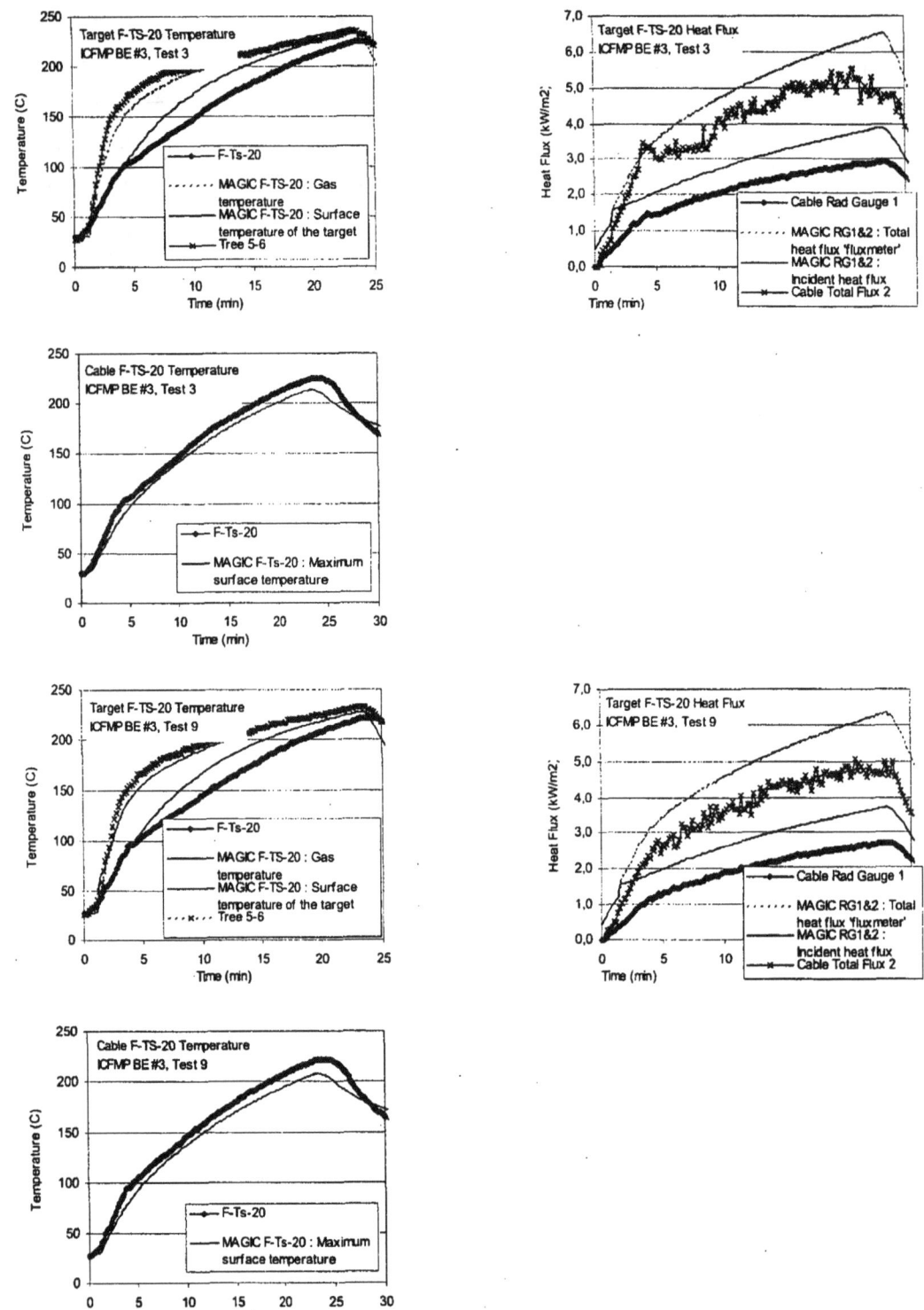

Figure A-57: Thermal Environment near Cable F, ICFMP BE #3, Tests 3 and 9

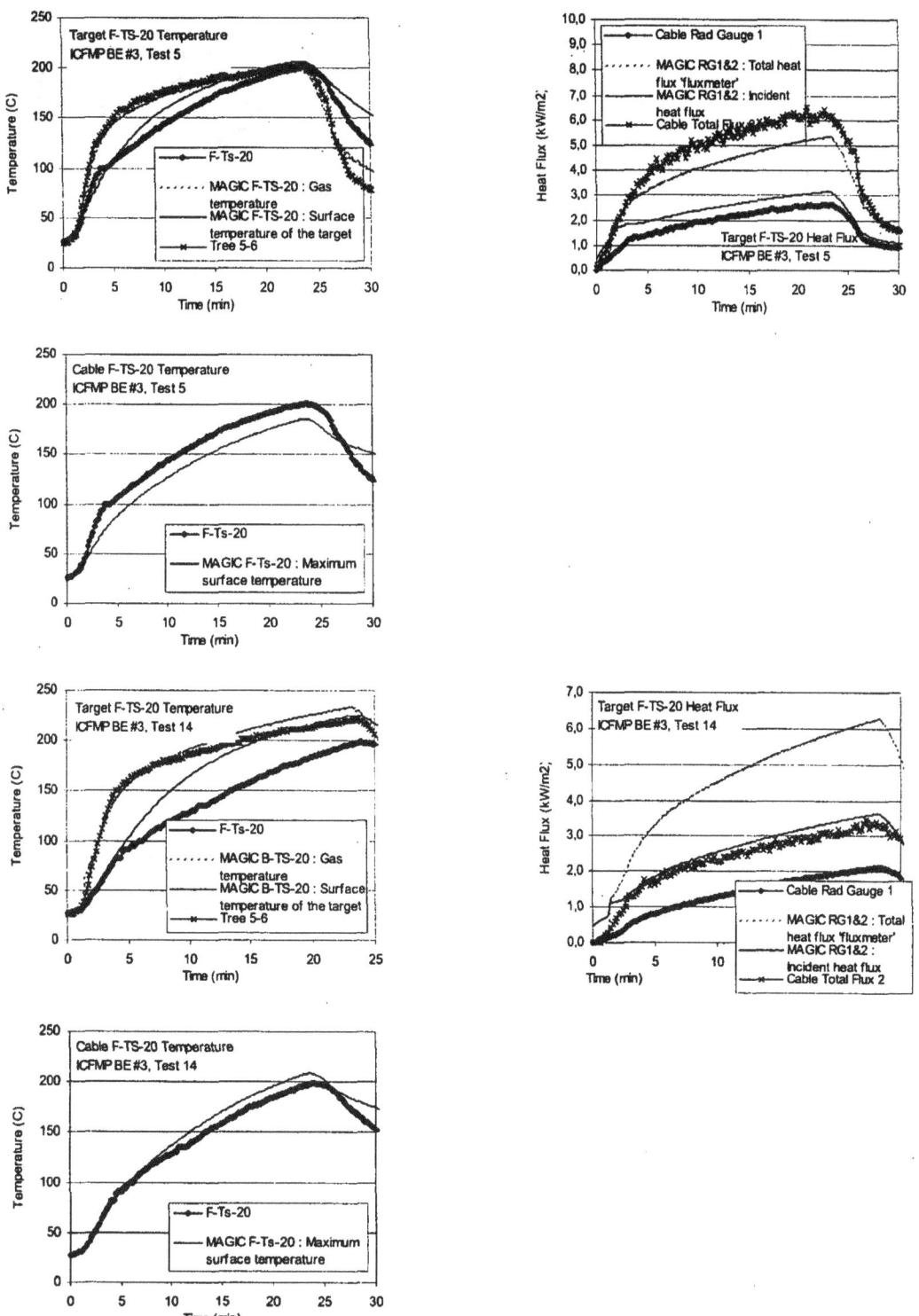

Figure A-58: Thermal Environment near Cable F, ICFMP BE #3, Tests 5 and 14

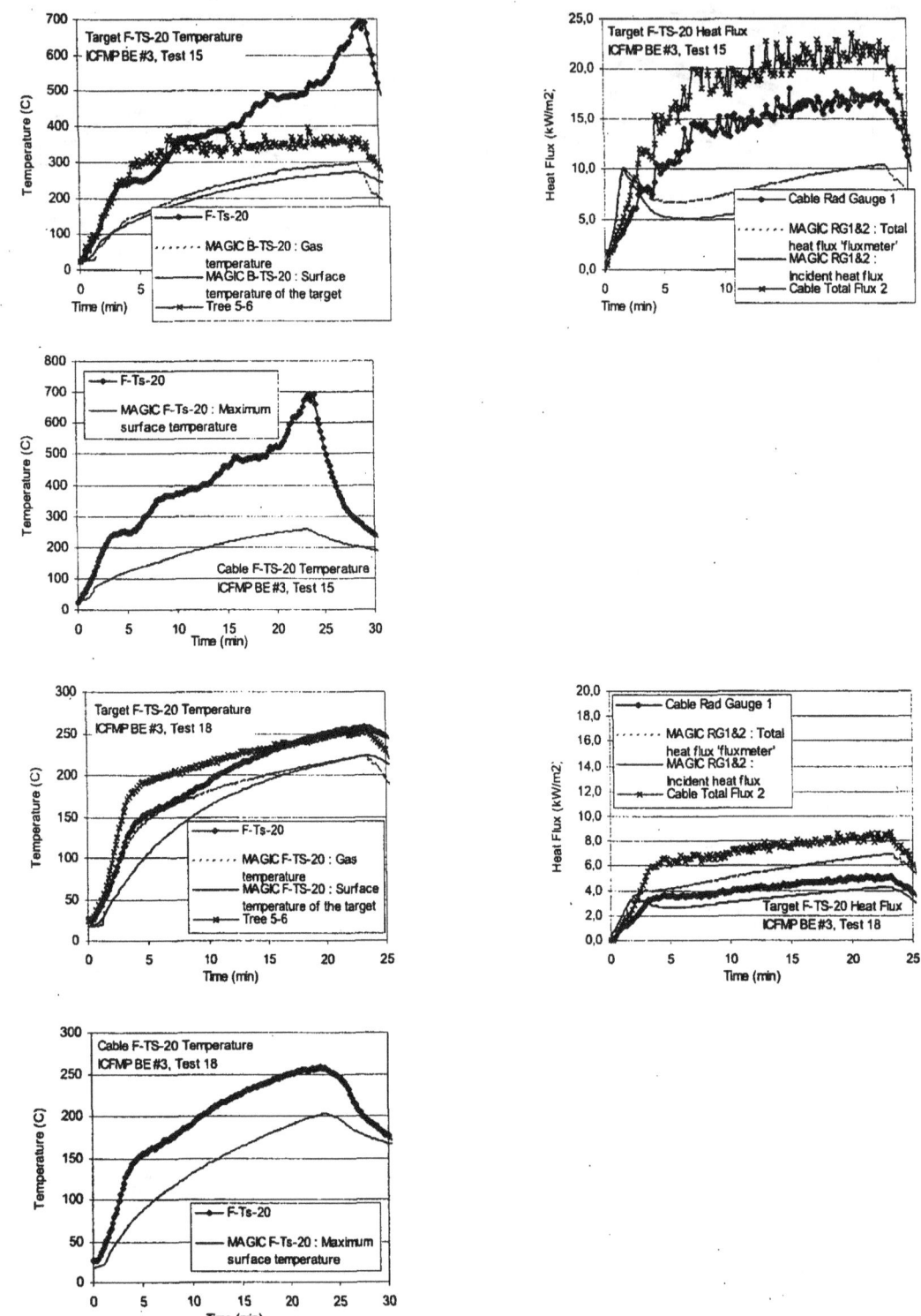

Figure A-59: Thermal Environment near Cable F, ICFMP BE #3, Tests 15 and 18

Table A-21: Relative Differences for Surface Temperature of Cable F

Power Cable	Target Surface Temp, F-TS-20			Cable Surface Temp, F-TS-20		
	ΔE (°C)	ΔM (°C)	Relative Diff	ΔE (°C)	ΔM (°C)	Relative Diff
Test 1	83	109	32%	83	95	15%
Test 7	90	109	21%	87	101	12%
Test 2	129	183	42%	129	150	16%
Test 8	131	181	38%	131	148	13%
Test 4	149	189	27%	149	158	6%
Test 10	150	194	29%	150	174	16%
Test 13	143	219	52%	143	188	31%
Test 16	168	227	35%	168	199	19%
Test 17						
Test 3	195	202	3%	195	182	-7%
Test 9	195	200	3%	195	180	-8%
Test 5	175	178	2%	175	157	-10%
Test 14	171	197	15%	171	179	5%
Test 15	669	245	-63%	669	231	-66%
Test 18	232	205	-11%	232	184	-20%

Table A-22: Relative Differences for Radiative and Total Heat Flux to Cable F

Power Cable	Radiant Heat Flux Gauge 1			Total Heat Flux, Gauge 2		
	ΔE (kW/m^2)	ΔM (kW/m^2)	Relative Diff	ΔE (kW/m^2)	ΔM (kW/m^2)	Relative Diff
Test 1	0.87	1.24	44%	1.60	3.05	90%
Test 7	1.99	3.60	81%	1.51	2.98	95%
Test 2	2.95	3.40	15%	4.77	6.89	44%
Test 8	2.02	3.33	65%	4.93	6.82	38%
Test 4	2.65	2.69	2%	5.02	6.49	29%
Test 10	0.82	1.22	48%	4.36	6.43	48%
Test 13	1.93	3.54	83%	7.28	11.40	56%
Test 16	1.93	3.29	71%	6.13	9.90	61%
Test 17	2.73	3.26	20%	1.85	3.43	83%
Test 3	2.90	6.91	139%	5.55	6.51	17%
Test 9	2.12	3.17	50%	5.08	6.33	25%
Test 5	18.29	9.59	-48%	6.45	5.34	-18%
Test 14	2.76	5.73	108%	3.46	6.26	77%
Test 15	0.88	1.49	69%	23.94	10.44	-56%
Test 18	5.18	3.87	-25%	8.74	6.94	-21%

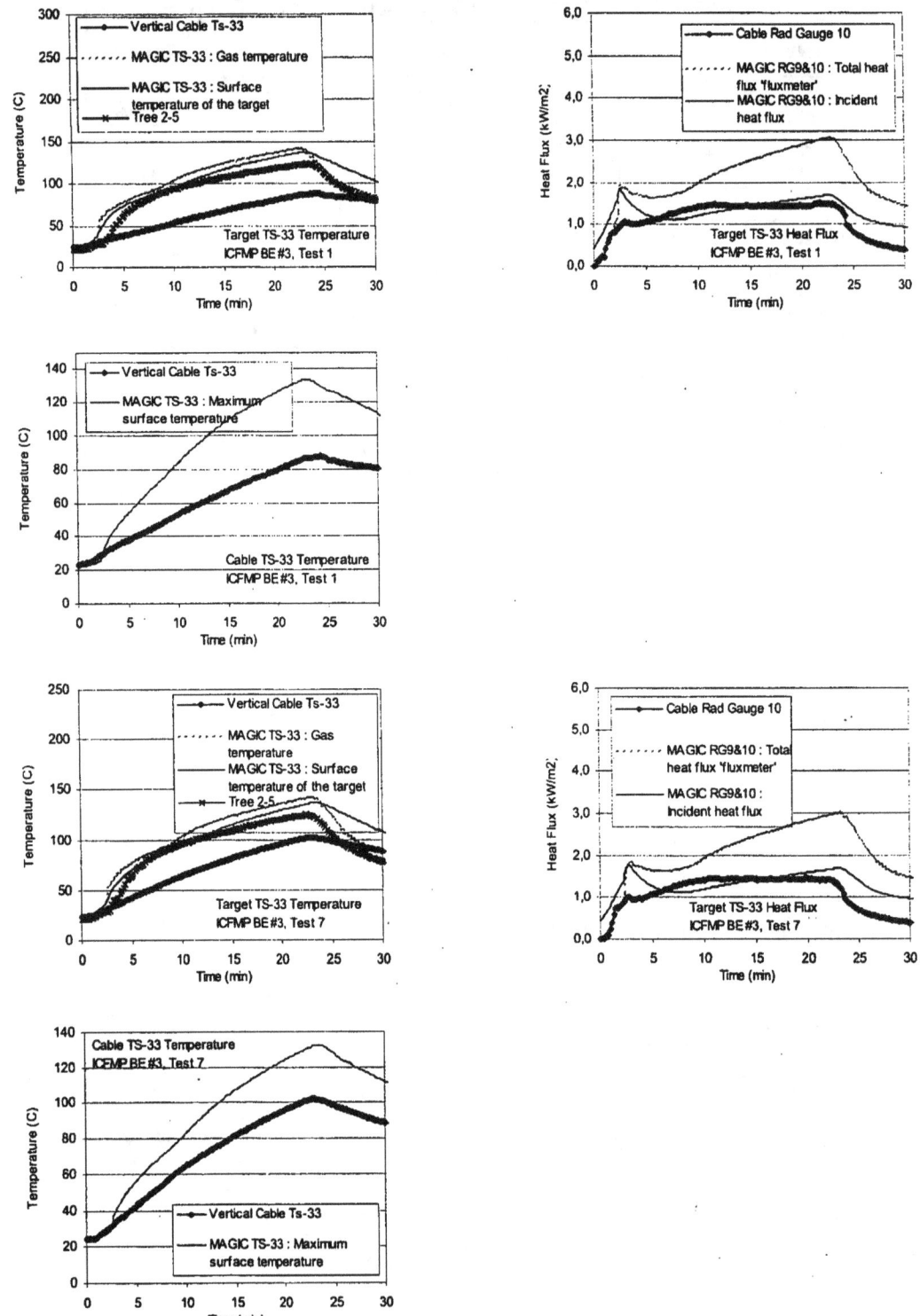

Figure A-60: Thermal Environment near Cable G, ICFMP BE #3, Tests 1 and 7

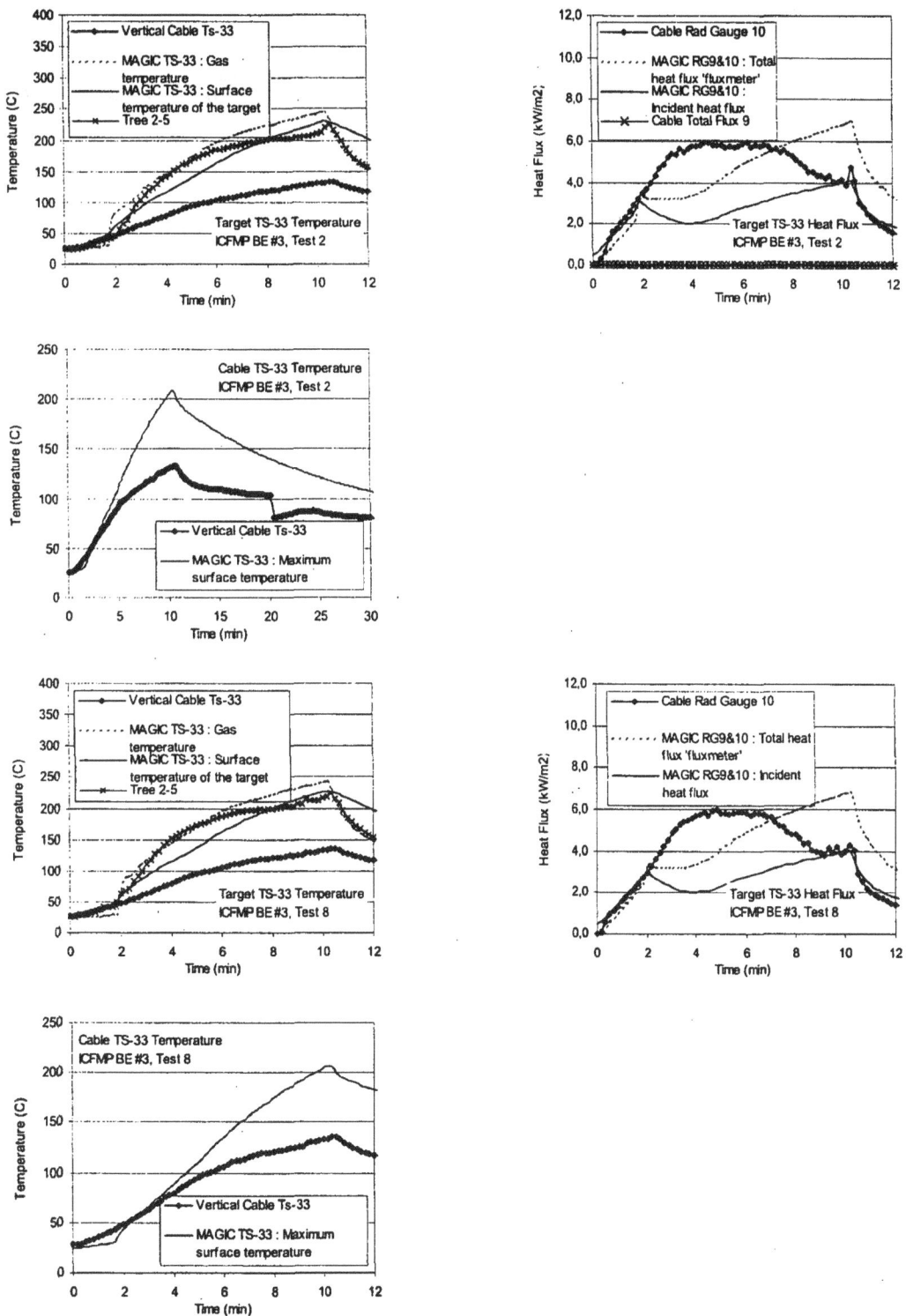

Figure A-61: Thermal Environment near Cable G, ICFMP BE #3, Tests 2 and 8

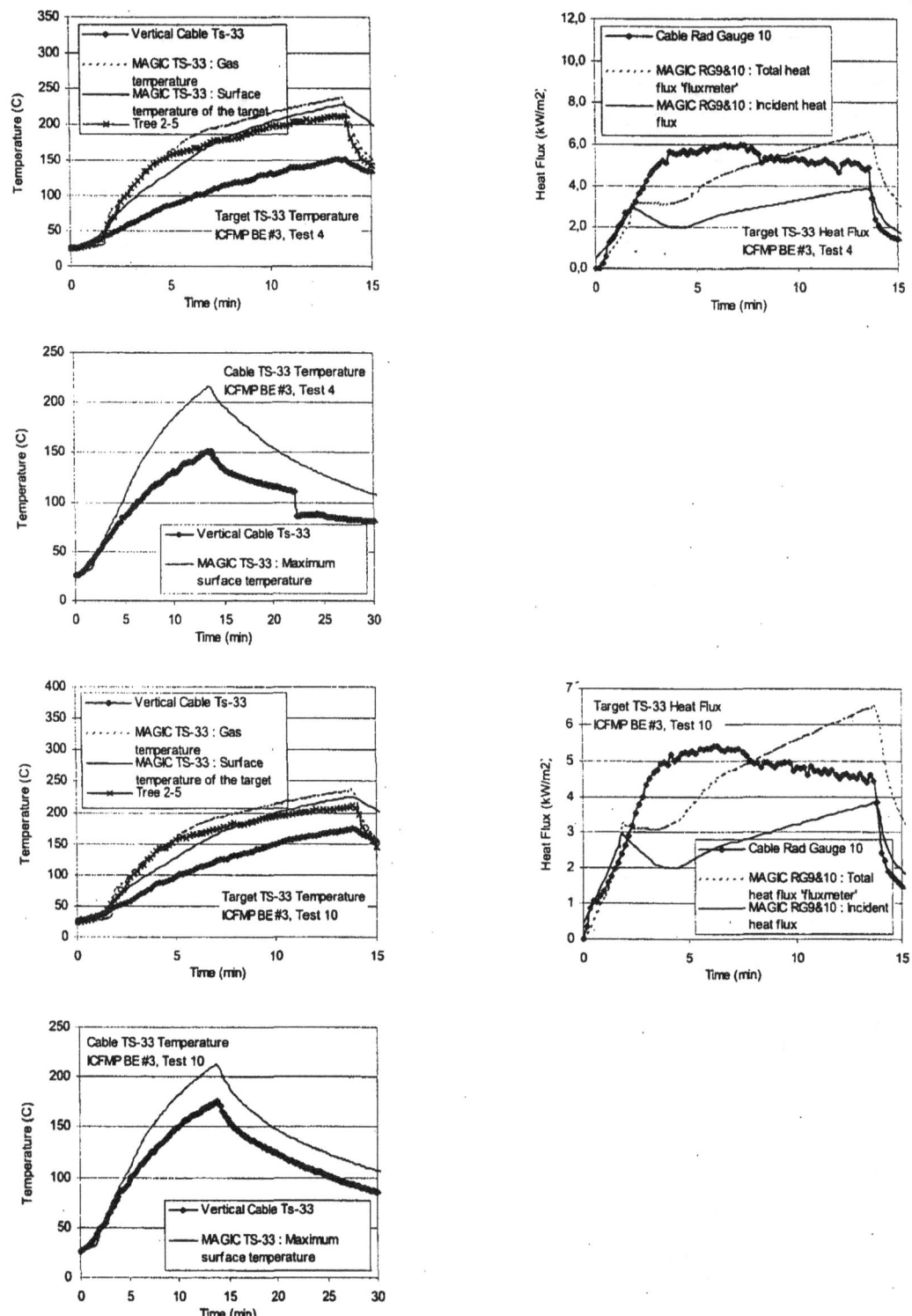

Figure A-62: Thermal Environment near Cable G, ICFMP BE #3, Tests 4 and 10

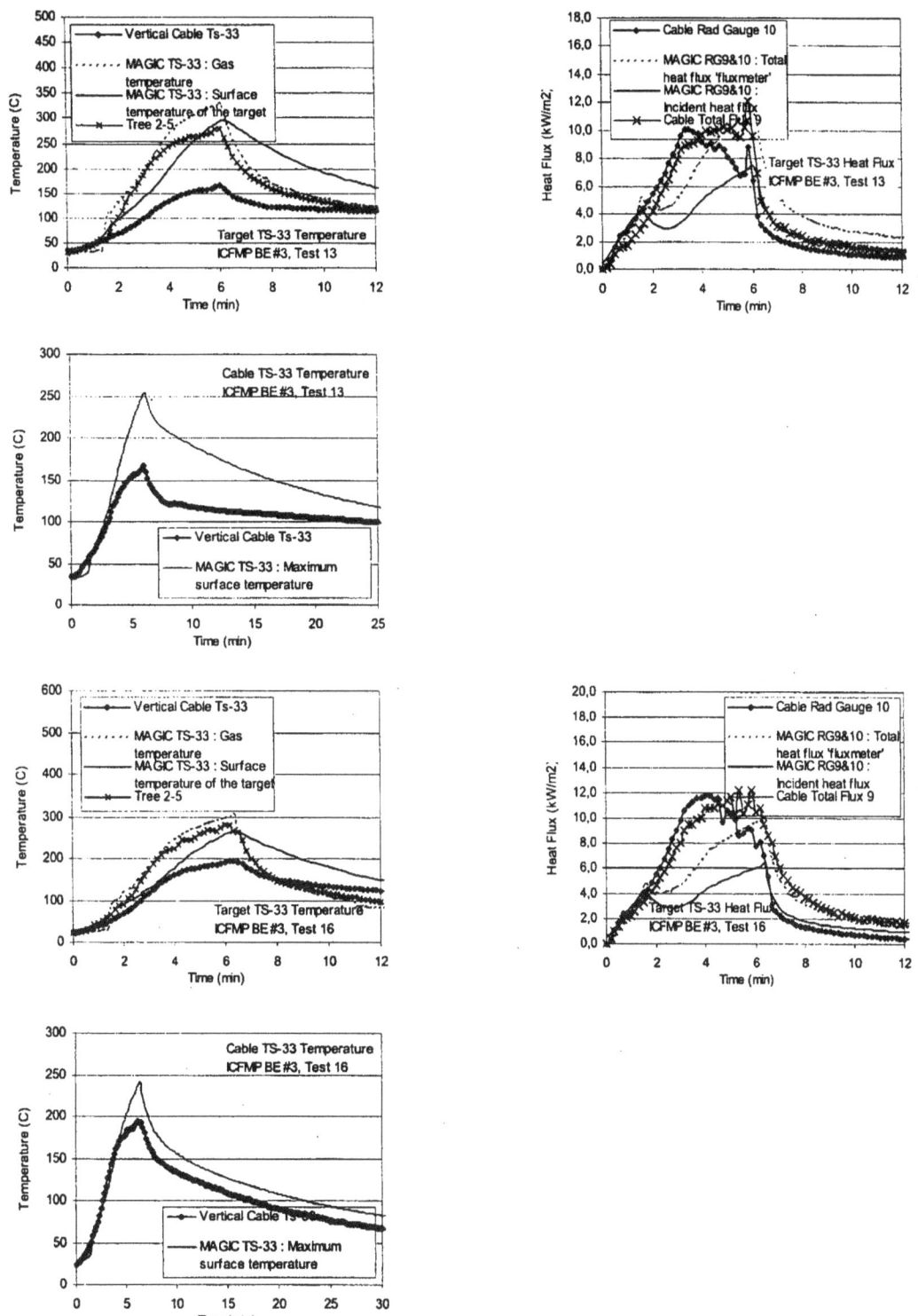

Figure A-63: Thermal Environment near Cable G, ICFMP BE #3, Tests 13 and 16

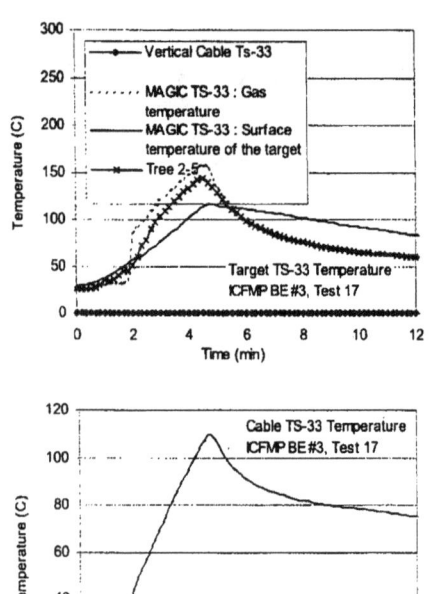

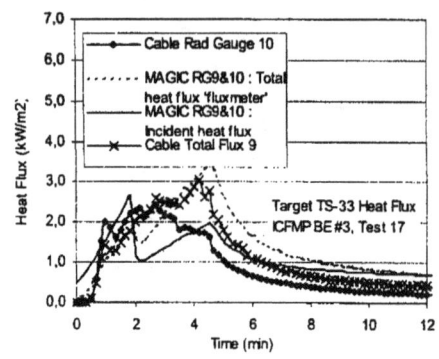

Figure A-64: Thermal Environment near Cable G, ICFMP BE #3, Test 17

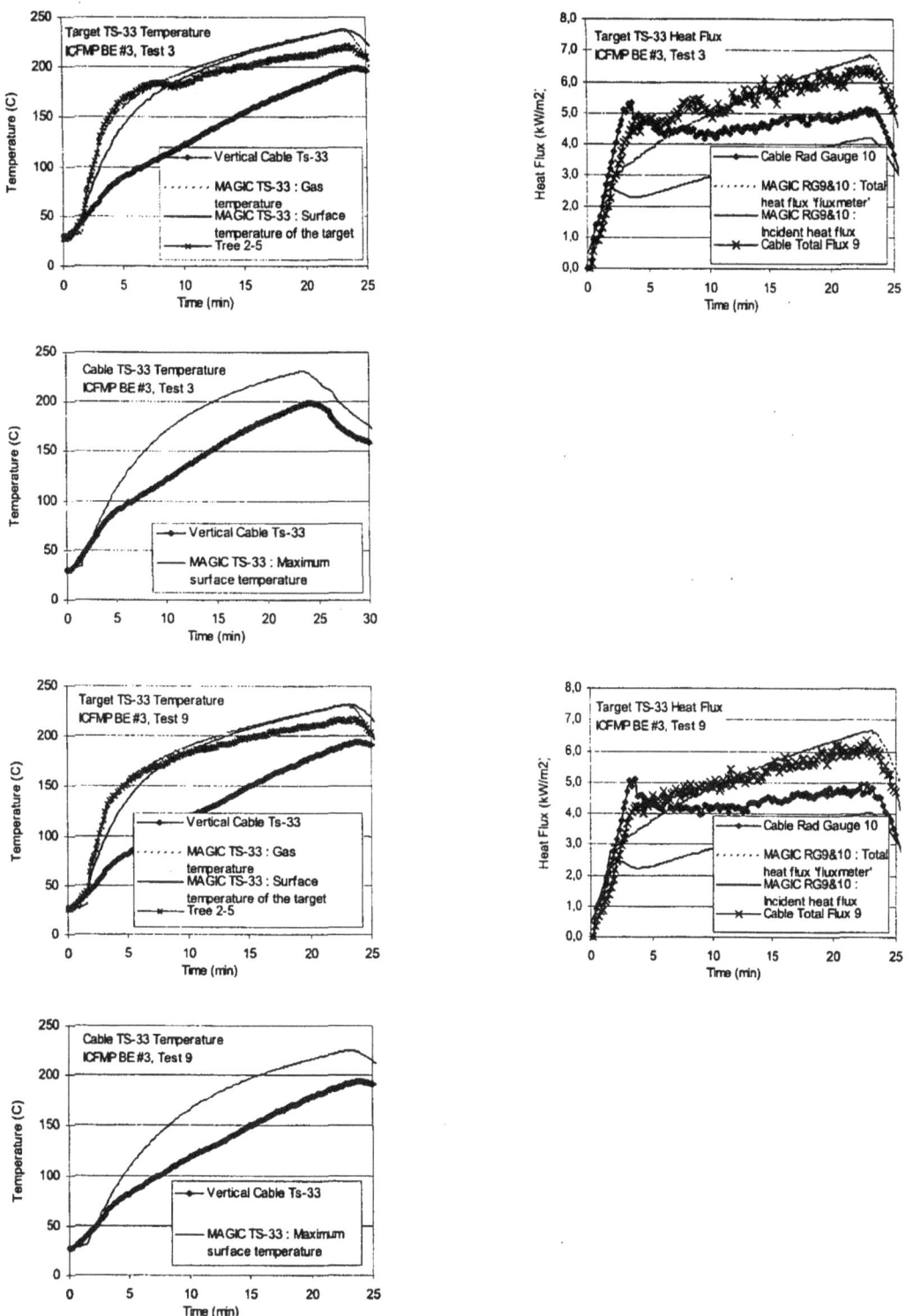

Figure A-65: Thermal Environment near Cable G, ICFMP BE #3, Tests 3 and 9

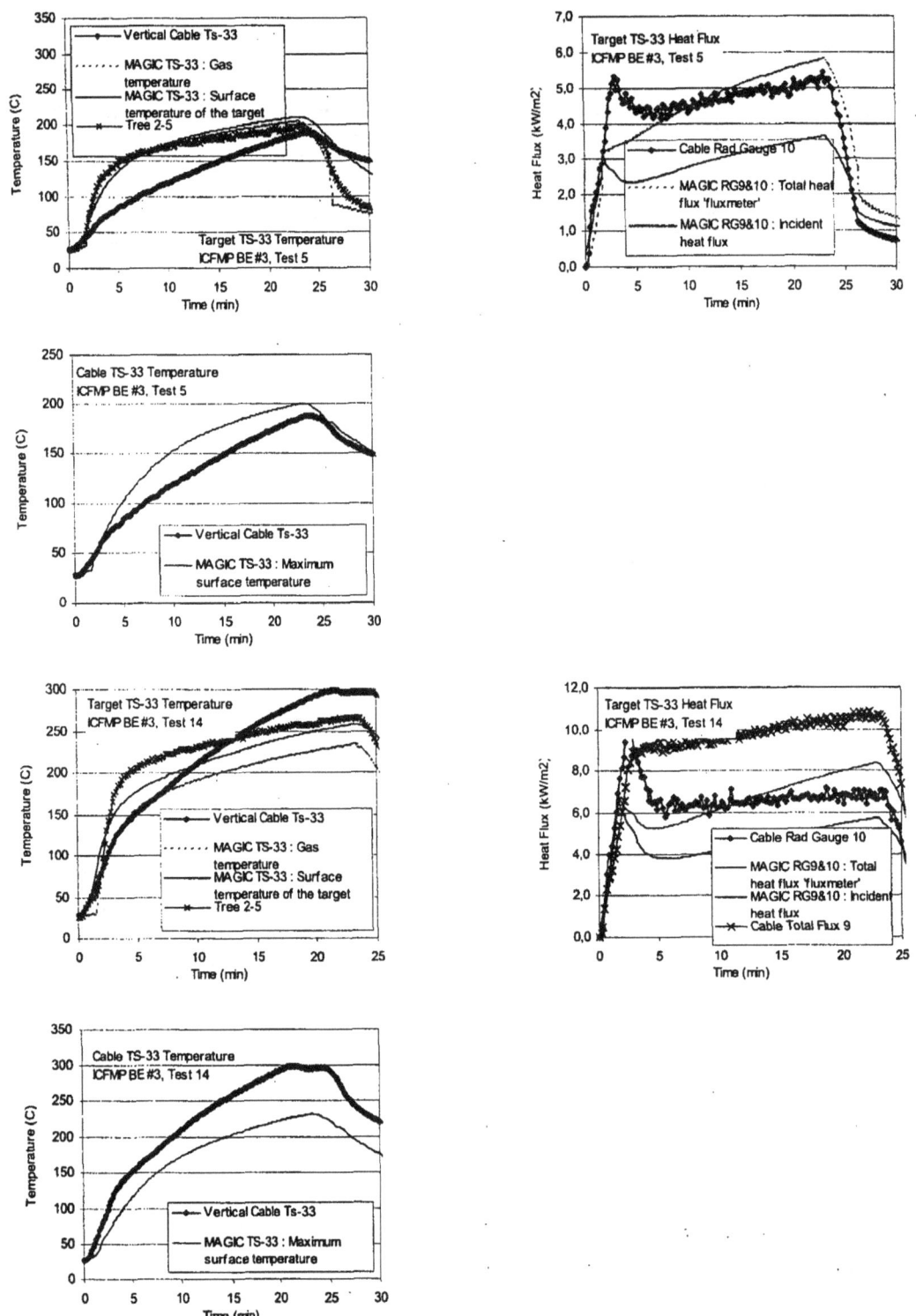

Figure A-66: Thermal Environment near Cable G, ICFMP BE #3, Tests 5 and 14

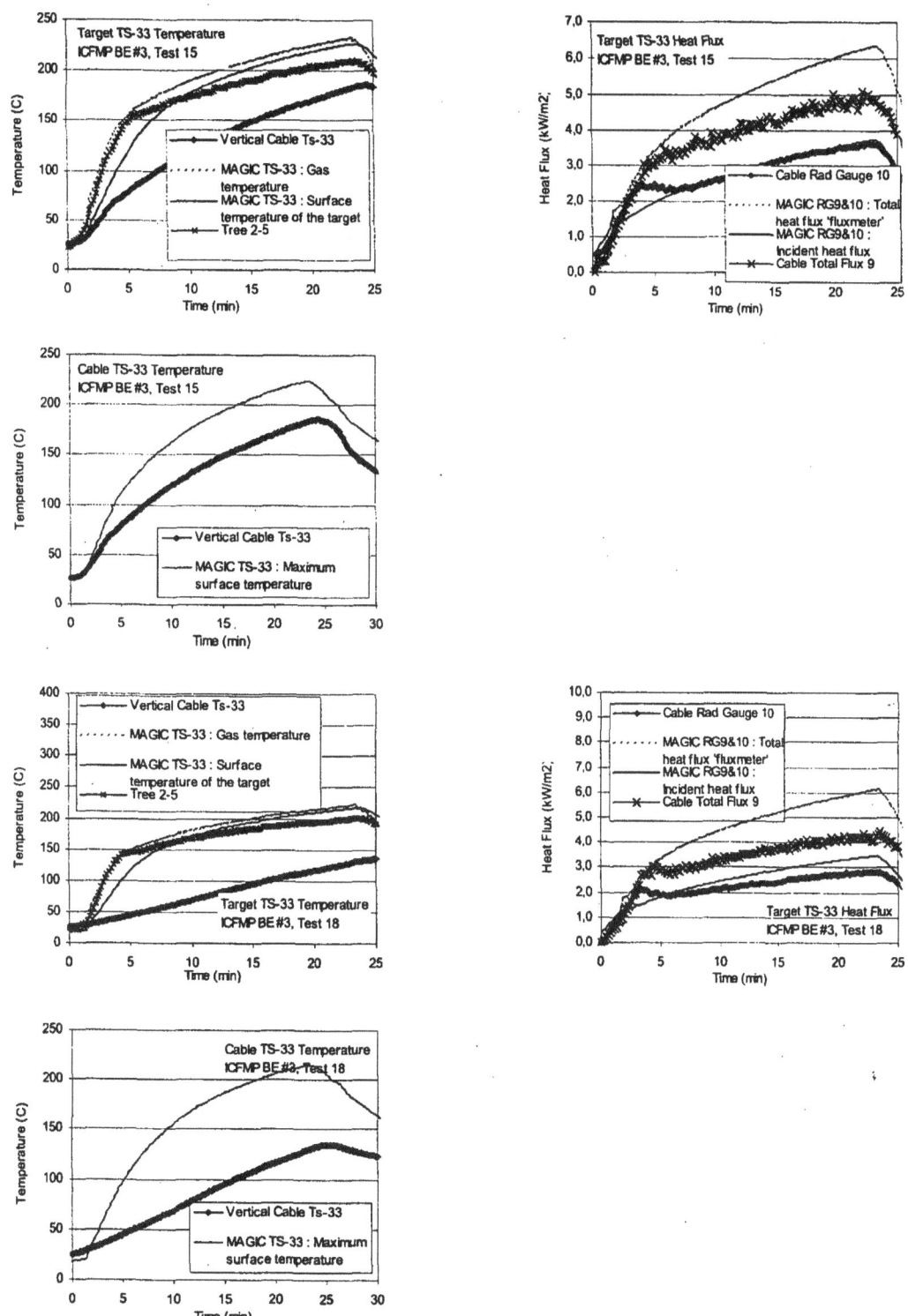

Figure A-67: Thermal Environment near Cable G, ICFMP BE #3, Tests 15 and 18

Table A-23: Relative Differences for Surface Temperature of Cable G

Control Cable	Target Surface Temp, G-TS-33			Cable Surface Temp, G-TS-33		
	ΔE (°C)	ΔM (°C)	Relative Diff	ΔE (°C)	ΔM (°C)	Relative Diff
Test 1	64	114	79%	64	111	74%
Test 7	78	111	42%	90	107	37%
Test 2	107	204	91%	107	184	72%
Test 8	107	203	90%	107	182	70%
Test 4	125	201	61%	125	189	51%
Test 10	148	198	34%	148	186	25%
Test 13	133	264	98%	133	222	66%
Test 16	169	243	44%	169	216	27%
Test 17						
Test 3	169	206	22%	169	200	19%
Test 9	165	204	23%	165	198	19%
Test 5	161	183	14%	161	172	7%
Test 14	270	230	-15%	270	204	-24%
Test 15	160	200	25%	160	197	23%
Test 18	106	200	89%	106	197	87%

Table A-24: Relative Differences for Radiative and Total Heat Flux to Cable G

Control Cable	Radiant Heat Flux Gauge 10			Total Heat Flux, Gauge 9		
	ΔE (kW/m^2)	ΔM (kW/m^2)	Relative Diff	ΔE (kW/m^2)	ΔM (kW/m^2)	Relative Diff
Test 1	1.51	1.41	-6%			
Test 7	5.97	3.64	-39%	1.89	2.99	56%
Test 2	5.36	3.71	-31%			
Test 8	6.00	3.41	-43%	5.98	6.86	15%
Test 4	5.45	3.16	-42%	6.42	6.58	2%
Test 10	1.47	1.33	-10%	6.20	6.51	5%
Test 13	6.03	3.59	-40%	12.18	11.50	-6%
Test 16	5.15	3.58	-31%	12.23	10.05	-18%
Test 17	5.42	3.37	-38%	3.07	3.43	11%
Test 3	10.06	7.01	-30%	6.45	6.82	6%
Test 9	10.50	6.12	-42%	6.37	6.65	4%
Test 5	3.73	3.28	-12%	6.69	5.80	-13%
Test 14	11.96	5.87	-51%	10.90	8.39	-23%
Test 15	2.42	2.15	-11%	5.12	6.36	24%
Test 18	2.85	3.06	8%	4.45	6.13	38%

A.8.2 ICFMP BE #4

Targets in BE #4, Test 1 were three material probes made of concrete, aerated concrete, and steel. Sensor M29 represents the aerated concrete material, while Sensors M33 and M34 represent the concrete and steel materials, respectively.

MAGIC appears to over-predict both total heat flux and surface temperature to the targets. The graphical comparisons for heat flux and surface temperature and the resulting relative differences are presented in Figure A-68 and Table A-17, respectively.

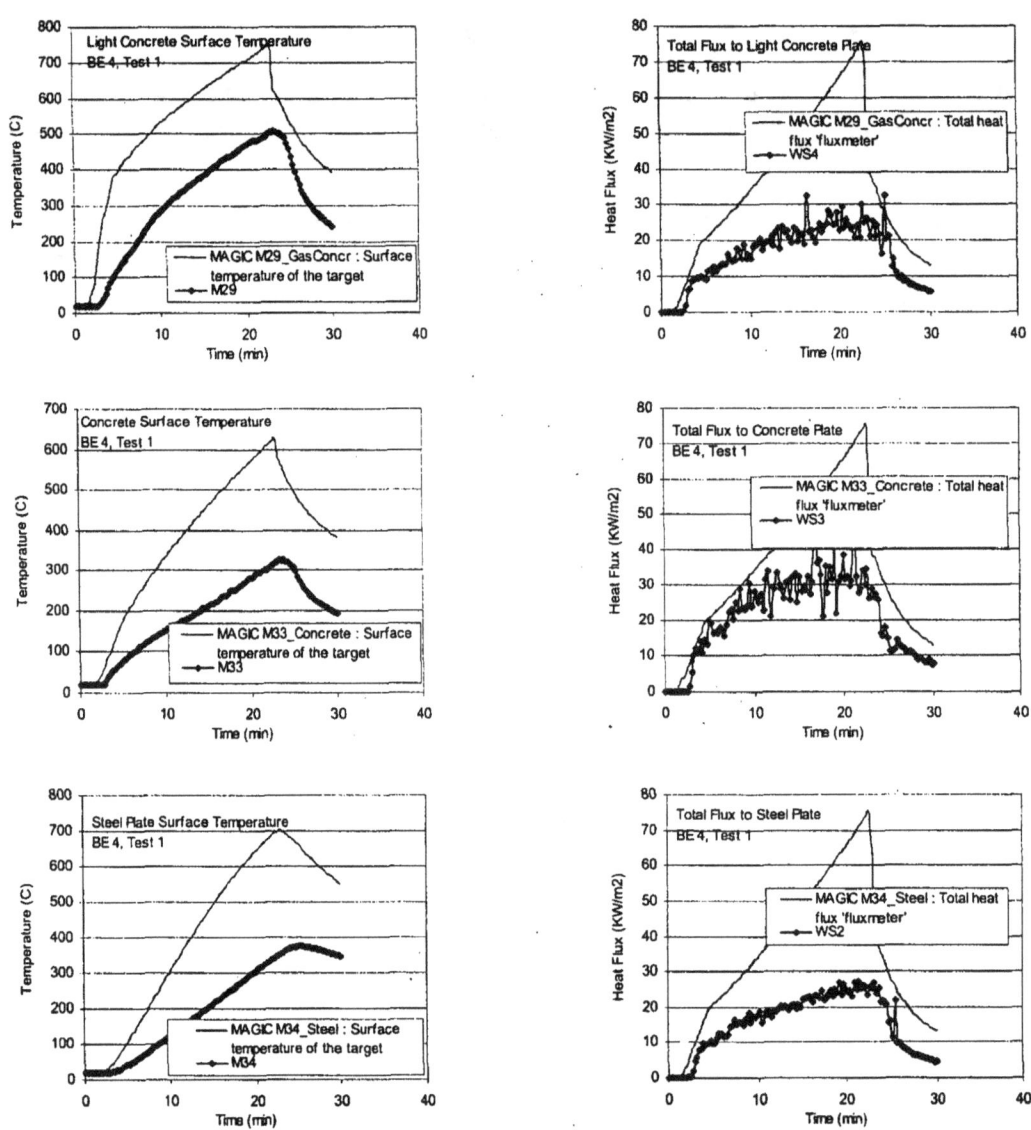

Figure A-68: Heat Flux and Surface Temperatures of Target Slabs, ICFMP BE #4, Test 1

Table A-25: Relative Differences for Surface Temperature and Total Heat Flux to Targets

		ΔE (°C)	ΔM (°C)	Relative Diff	ΔE (kW/m²)	ΔM (kW/m²)	Relative Diff
ICFMP 4-1	Steel, M34	356	684	92%	27.18	75.65	178%
	Concrete, M33	308	608	97%	46.56	75.65	63%
	Gas Concrete, M29	489	728	49%	32.41	75.65	133%

A.8.3 ICFMP BE #5

A vertical cable tray was positioned near a wall opposite the fire. Heat flux gauges were inserted in between two bundles of cables, one containing power cables, and the other containing control cables. The following pages show plots of the gas temperatures, heat fluxes, and cable surface temperatures at three vertical locations along the tray.

Figure A-69 compiles the graphical comparisons for total heat flux and surface temperature. Table A-26 and Table A-27 list the corresponding relative differences.

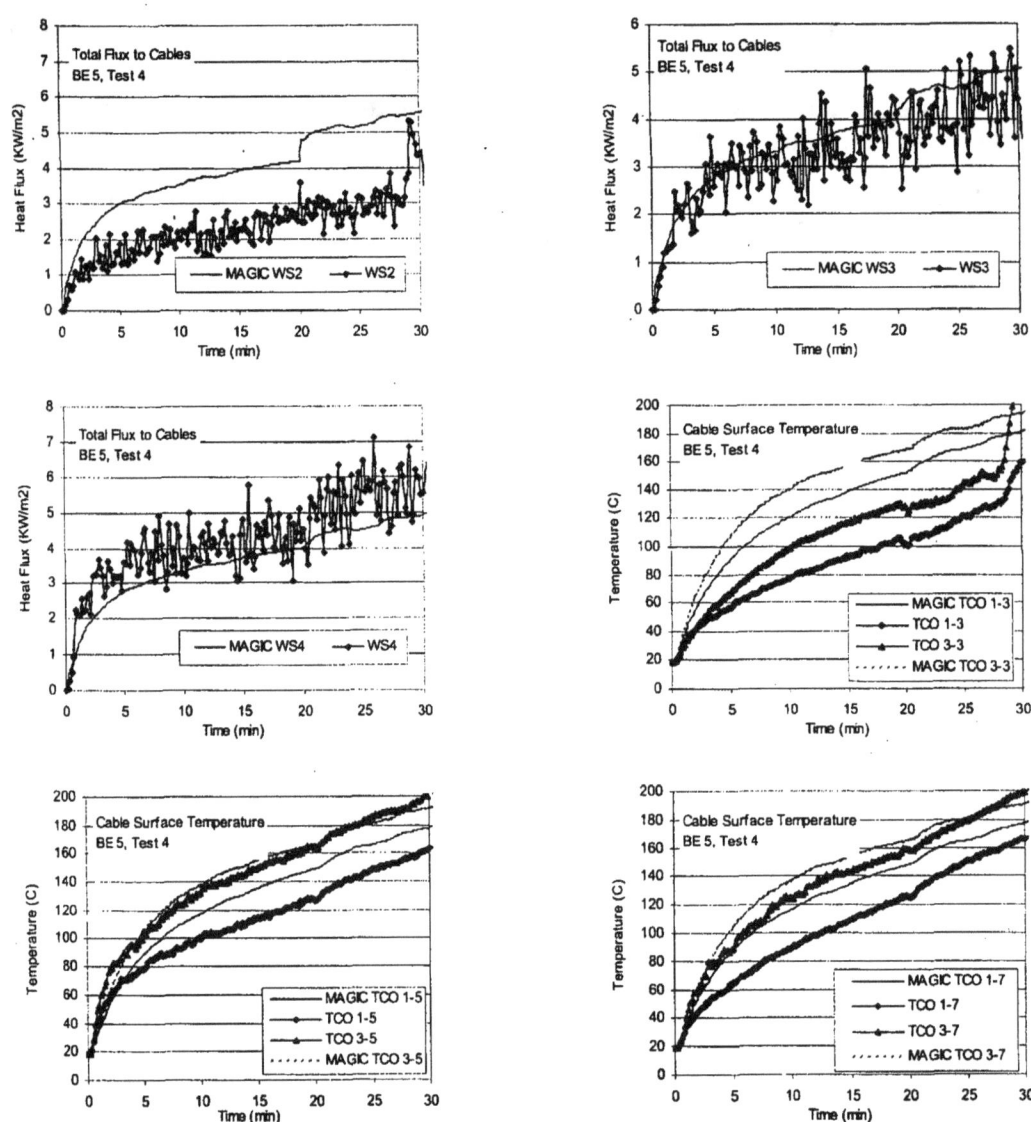

Figure A-69: Thermal Environment near Vertical Cable Tray, ICFMP BE #5, Test 4

Table A-26: Relative Differences for Surface Temperature

Surface Temperature	Instrument	ΔE (°C)	ΔM (°C)	Relative Difference
ICFMP 5-4	TCO 1-3	141.1	165.1	17%
	TCO 3-3	144.3	165.2	14%
	TCO 1-5	147.8	159.2	8%
	TCO 3-5	222.5	159.3	-28%
	TCO 1-7	182.6	157.6	-14%
	TCO 3-7	180.2	157.7	-13%

Table A-27: Relative Differences for Total Heat Flux

Total Heat Flux	Instrument	ΔE (kW/m^2)	ΔM (kW/m^2)	Relative Diff
ICFMP 5-4	WS2	141	161	14%
	WS3	144	174	21%
	WS4	148	158	7%

A.9 Wall Heat Flux and Temperature

Heat fluxes and surfaces temperatures at compartment walls, floor, and ceiling are available from ICFMP BE #3 and #5. This category is similar to that of the previous section, "Heat Flux and Surface Temperature of Targets," with the exception that the focus here is on the compartment walls, floor, and ceiling.

MAGIC offers two alternatives for wall temperature and heat flux results. The first alternative results from a heat balance at the surface of the wall in both the upper and lower layers. The second option is to locate a target characterized by the wall properties. The second option is preferred for validation purposes because the target can be placed in the same location as the experimental sensors.

A.9.1 ICFMP BE #3

Thirty-six heat flux gauges were positioned at various locations on all four compartment walls, plus the floor and ceiling. Comparisons between measured and predicted heat fluxes and surface temperatures are shown on the following pages for a selected number of locations. Over half of the measurement points were in roughly the same relative location to the fire and, hence, the measurements and predictions were similar. For this reason, data for the east and north walls are shown because the data from the south and west walls are comparable. Data from the south wall is used in cases where the corresponding instrument on the north wall failed, or the fire was positioned close to the south wall.

The heat flux gauges used on the compartment walls measured the *net,* rather than total, heat flux. In MAGIC, this measured heat flux is compared with the Target/Heat Flux/ Total Absorbed Heat Flux output option.

The following graphical comparisons are grouped per room surface (long wall, short wall, ceiling, or floor). The term "long wall" refers to either the north or south wall. The term "short wall" refers to either the east or west wall. Two sensors have been selected for comparison for each surface. Comparisons include both surface temperature and heat flux. The corresponding relative differences are provided after the graphical comparison for each room surface.

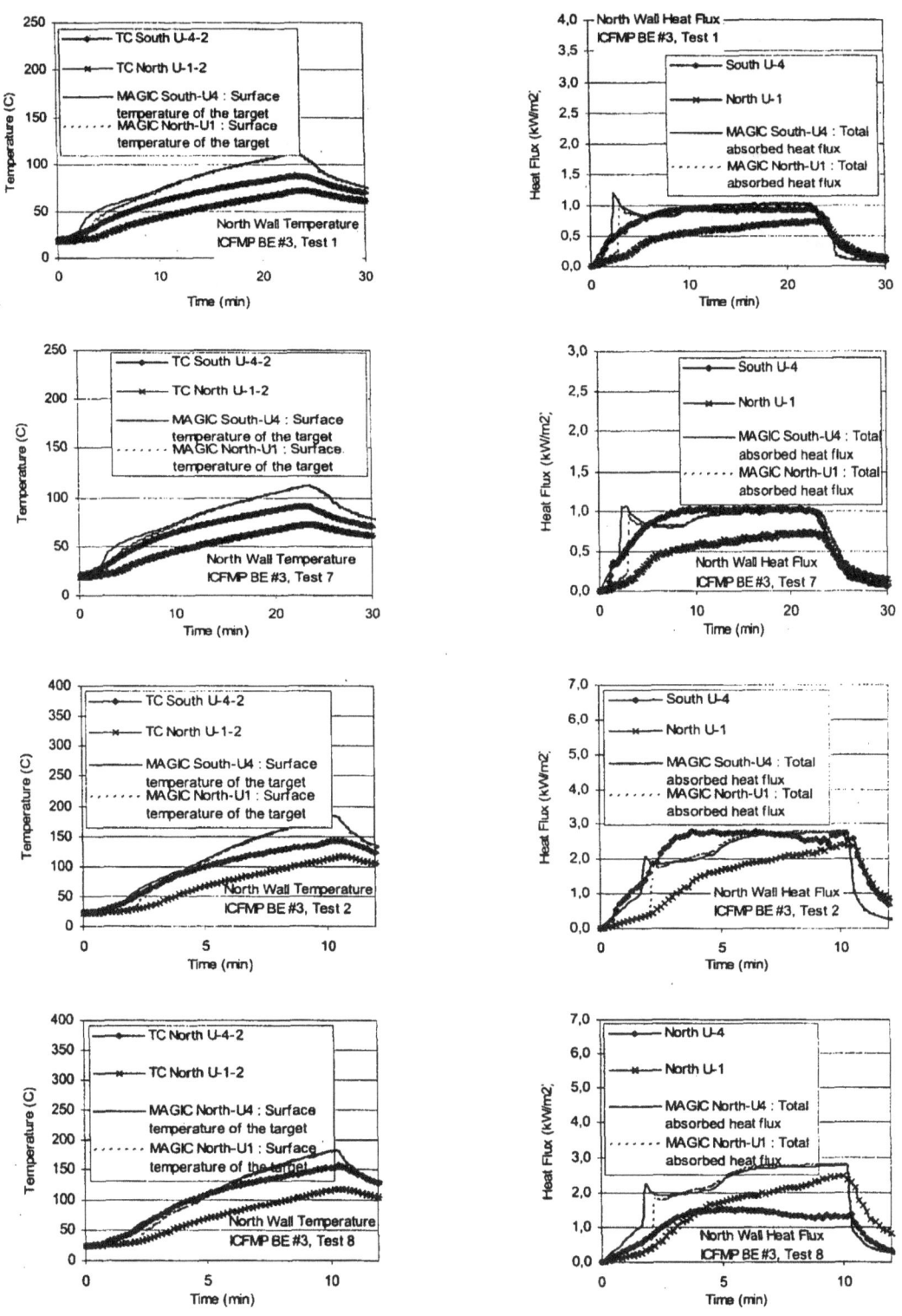

Figure A-70: Long-Wall Heat Flux and Surface Temperature, ICFMP BE #3, Closed-Door Tests

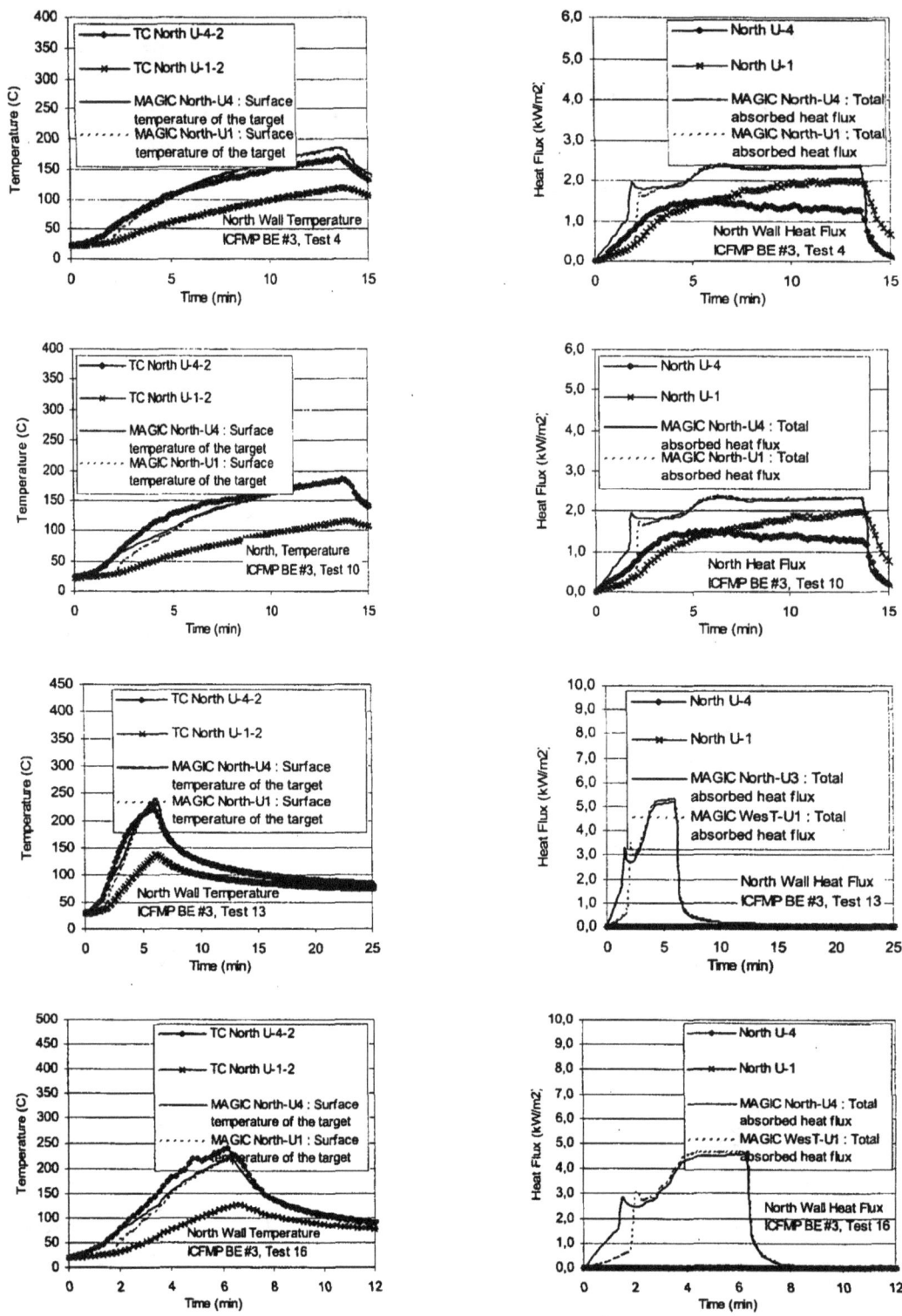

Figure A-71: Long-Wall Heat Flux and Surface Temperature, ICFMP BE #3, Closed-Door Tests

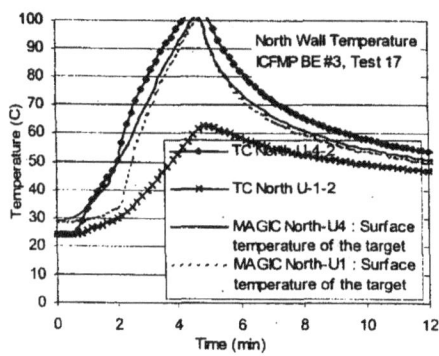

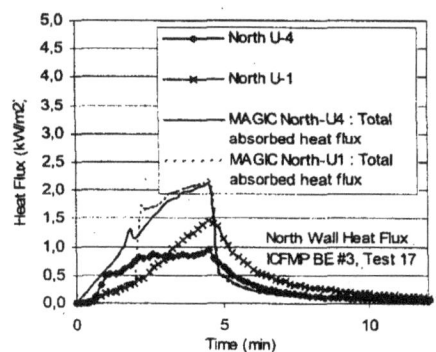

Open-door tests

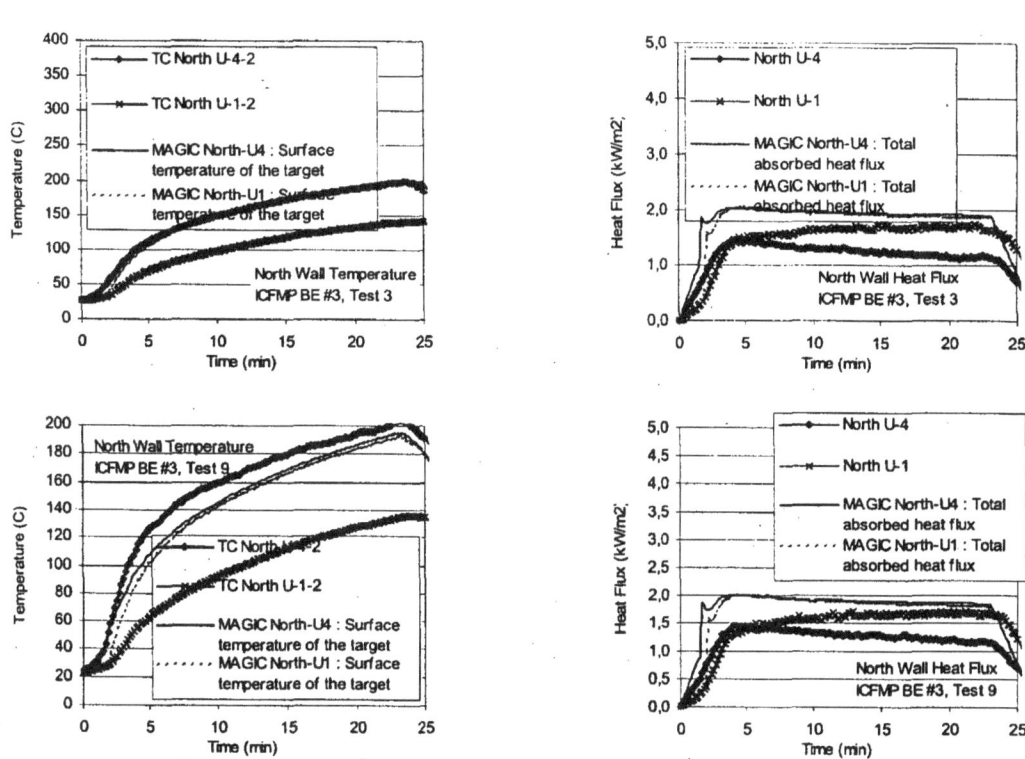

Figure A-72: Long-Wall Heat Flux and Surface Temperature, ICFMP BE #3, Closed- and Open-Door Tests

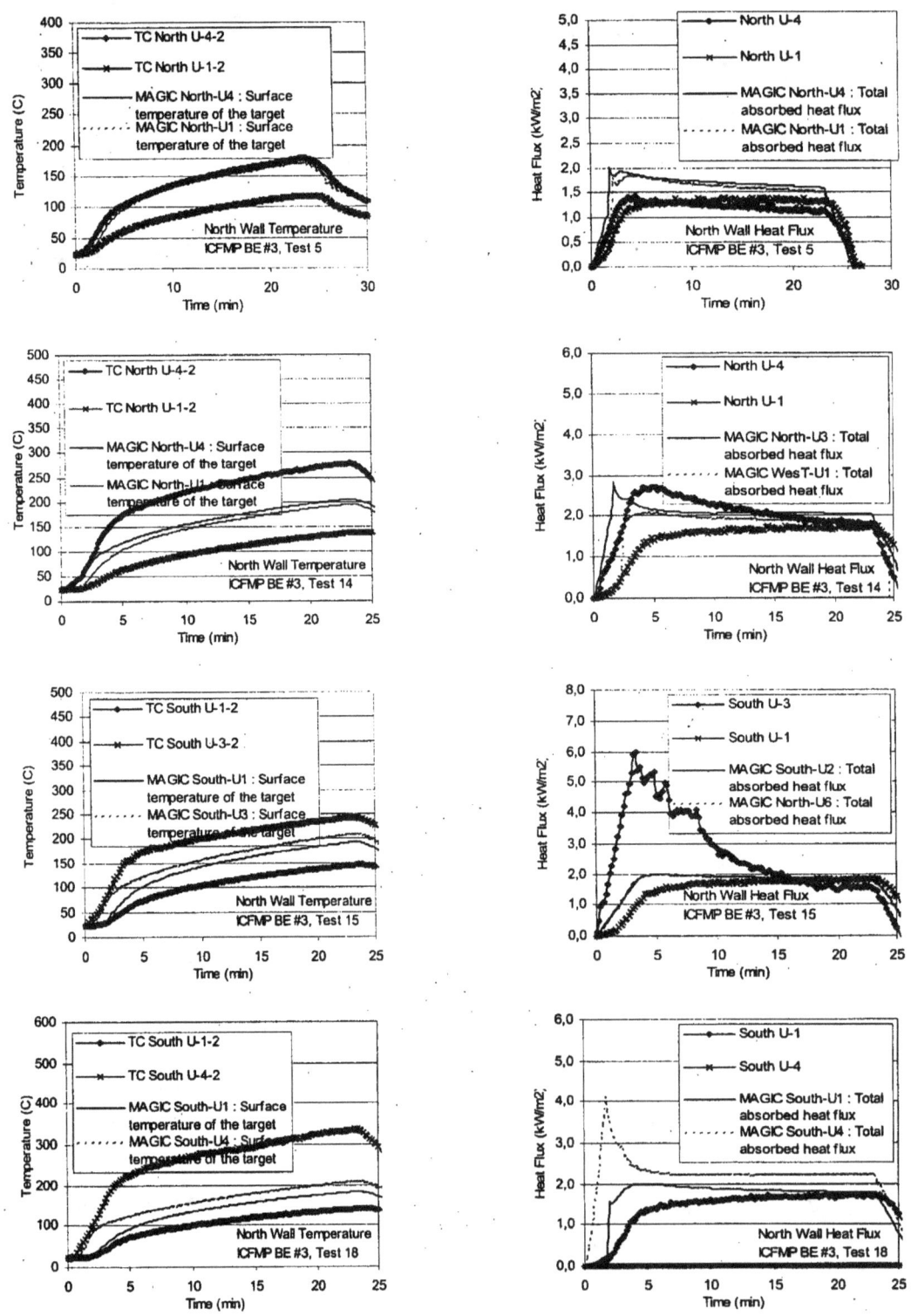

Figure A-73: Long-Wall Heat Flux and Surface Temperature, ICFMP BE #3, Open-Door Tests

Table A-28: Relative Differences for Temperature and Total Heat Flux Corresponding to the Long Wall

Long Wall	Instrument	Wall Temperature			Wall Total Heat Flux		
		ΔE (°C)	ΔM (°C)	Relative Diff	ΔE (kW/m^2)	ΔM (kW/m^2)	Relative Diff
Test 1	N1	54	89	65%	0.7	1.1	49%
	S4	68	89	31%	1.0	1.2	25%
Test 7	N1	53	87	63%	0.8	1.0	29%
	S4	70	87	23%	1.1	1.1	-5%
Test 2	N1	96	158	65%	2.4	2.8	17%
	S4	120	158	32%	2.8	2.8	-1%
Test 8	N1	95	157	66%	2.5	2.8	12%
	N4	120	157	19%	1.5	2.8	83%
Test 4	N1	97	159	64%	2.0	2.4	20%
	N4	124	159	9%	1.5	2.3	59%
Test 10	N1	94	158	68%	2.0	2.4	20%
	N4	124	158	-3%	1.5	2.3	53%
Test 13	N1	110	209	91%			
	N4	151	210	5%			
Test 16	N1	107	195	83%			
	N4	150	196	-10%			
Test 17	N1	39	70	79%	1.5	2.1	47%
	N4	54	71	-13%	0.9	2.1	125%
Test 3	N1	114	168	47%	1.7	2.0	18%
	N4	169	170	-1%	1.4	2.0	40%
Test 9	N1	113	165	47%	1.7	2.0	17%
	N4	167	168	-6%	1.4	2.0	40%
Test 5	N1	94	143	52%	1.5	1.8	20%
	N4	150	147	-5%	1.7	2.0	20%
Test 14	N1	114	167	46%	1.8	2.0	16%
	N4	146	176	-31%	2.7	2.6	-4%
Test 15	S1	115	167	34%	1.9	2.0	9%
	S3	308	182	-17%	5.9	3.1	-46%
Test 18	S1	109	166	41%	1.8	2.0	14%
	S4	312	191	-39%			

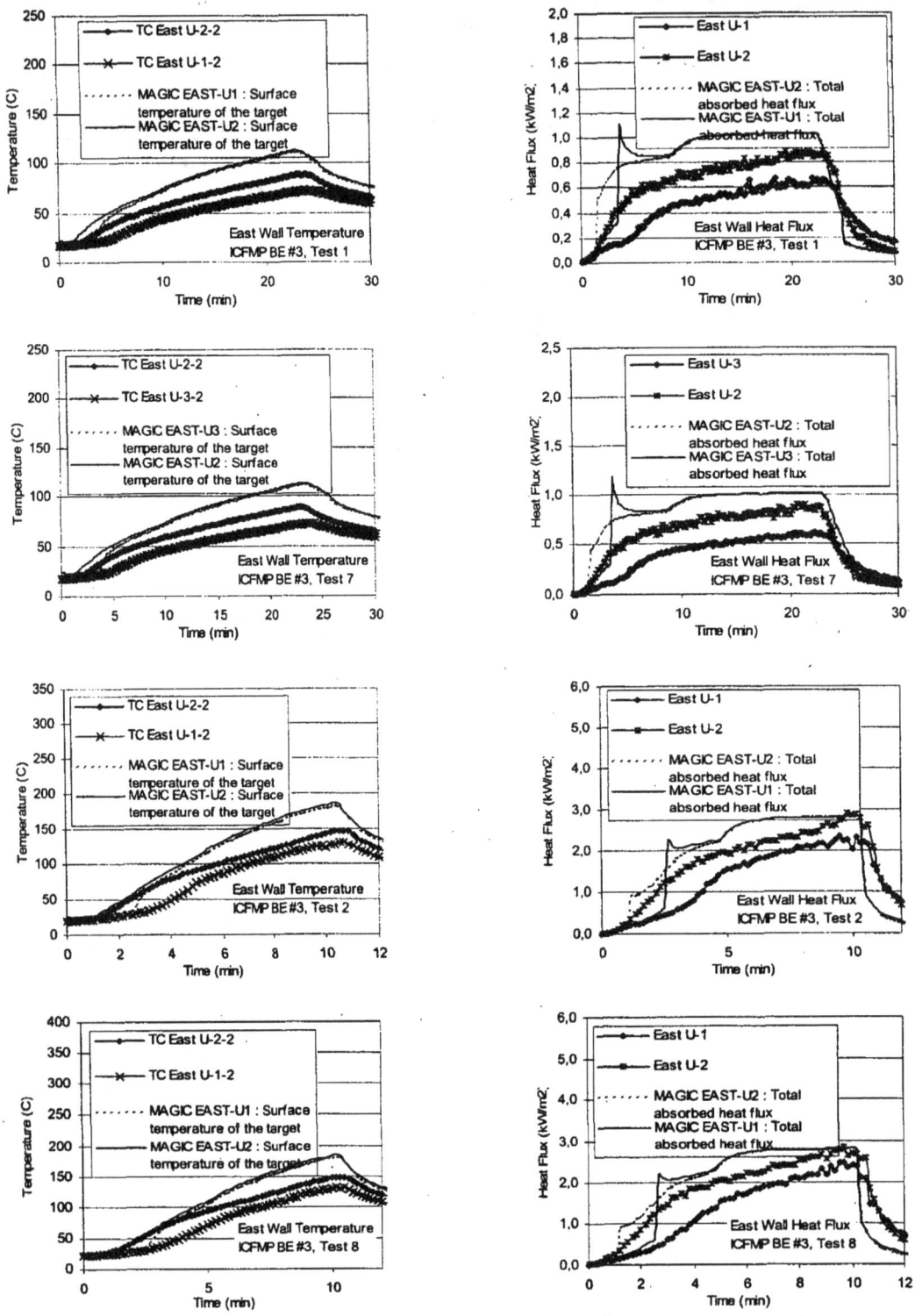

Figure A-74: Short-Wall Heat Flux and Surface Temperature, ICFMP BE #3, Closed-Door Tests

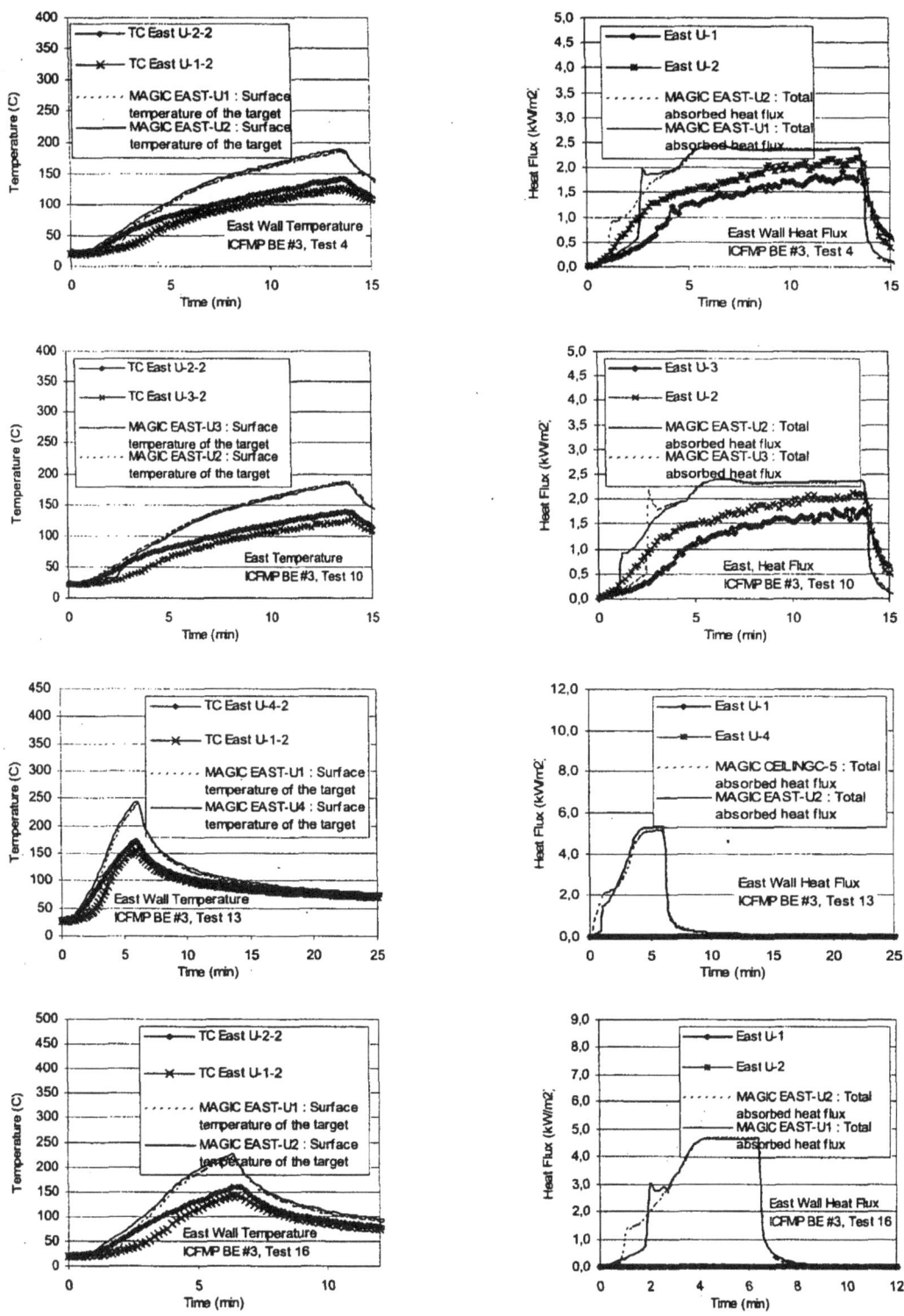

Figure A-75: Short-Wall Heat Flux and Surface Temperature, ICFMP BE #3, Closed-Door Tests

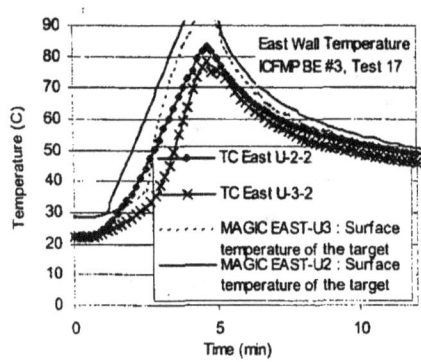 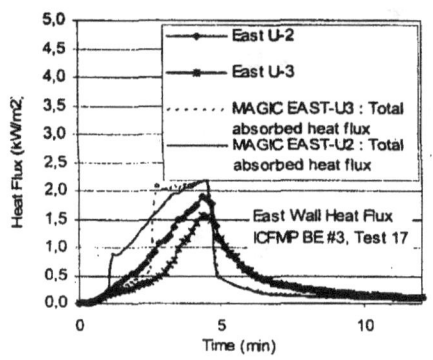

Open-door test

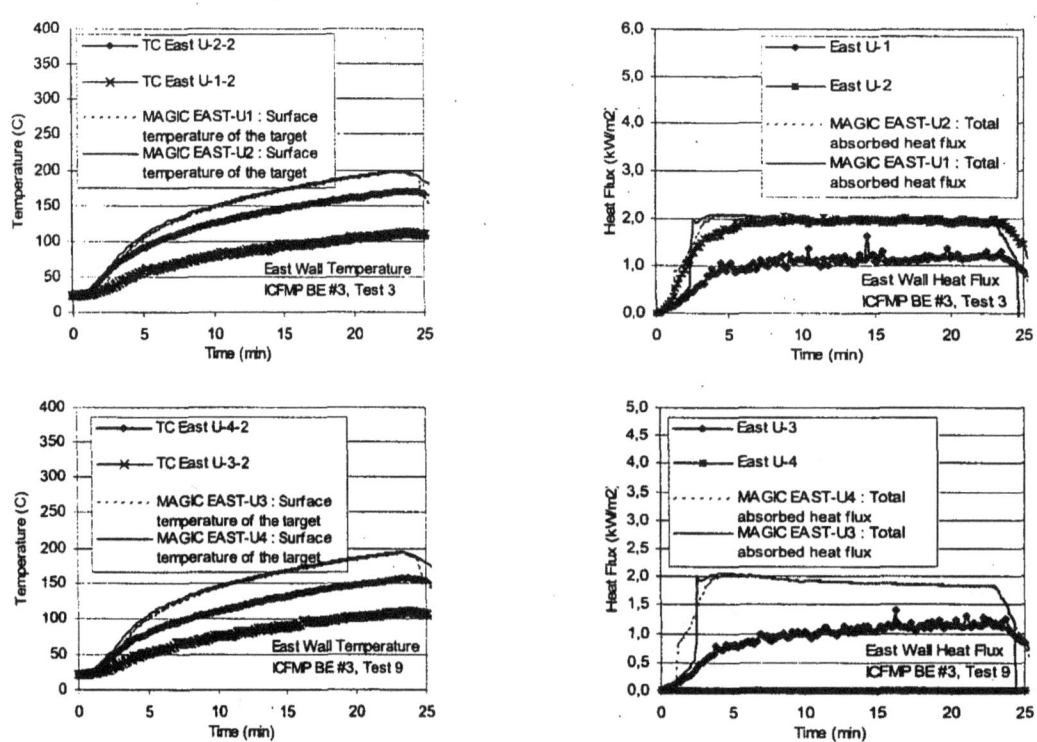

Figure A-76: Short-Wall Heat Flux and Surface Temperature, ICFMP BE #3, Closed- and Open-Door Tests

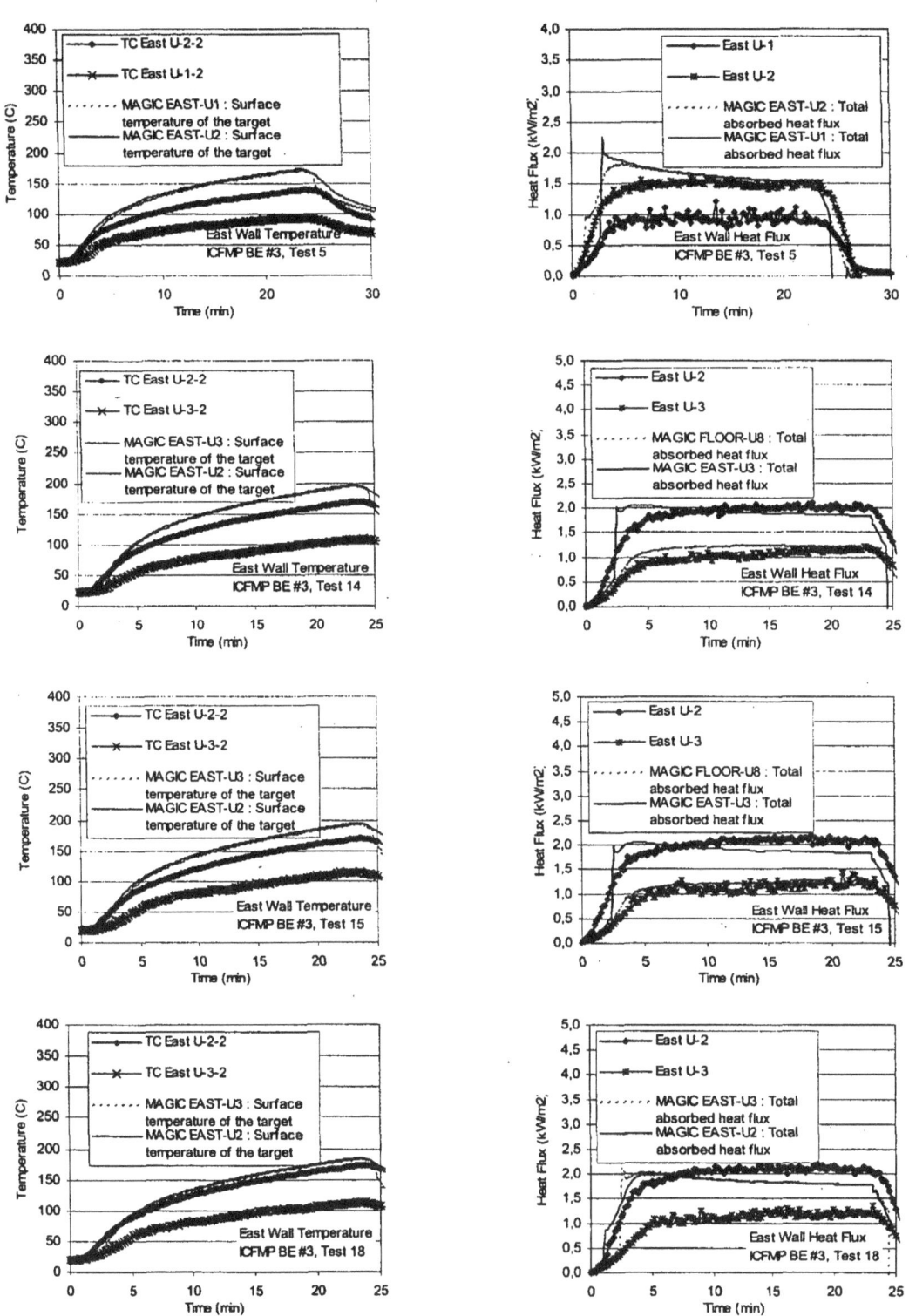

Figure A-77: Short-Wall Heat Flux and Surface Temperature, ICFMP BE #3, Open-Door Tests

Table A-29: Relative Differences for Temperature and Total Heat Flux Corresponding to the Short Wall

Short Wall	Instrument	Wall Temperature			Wall Total Heat Flux		
		ΔE (°C)	ΔM (°C)	Relative Diff	ΔE (kW/m²)	ΔM (kW/m²)	Relative Diff
	E1	55	89	61%	0.7	1.1	64%
Test 1	E2	71	91	28%	0.9	1.0	17%
	E3	53	87	59%	0.7	1.2	77%
Test 7	E2	70	88	26%	1.0	1.0	6%
	E1	110	158	44%	2.3	2.8	20%
Test 2	E2	125	162	29%	2.9	2.8	-2%
	E1	109	157	44%	2.5	2.8	12%
Test 8	E2	125	160	29%	2.9	2.8	-3%
	E1	106	159	50%	1.9	2.4	26%
Test 4	E2	121	162	34%	2.2	2.4	12%
	E3	102	158	51%	1.8	2.4	36%
Test 10	E2	117	161	38%	2.1	2.4	14%
	E1	127	208	65%			
Test 13	E4	55	214	290%			
	E1	123	195	58%			
Test 16	E2	141	201	42%			
	E3	52	69	22%	1.6	2.2	40%
Test 17	E2	61	74	21%	1.9	2.2	14%
	E1	87	168	92%	1.6	2.1	27%
Test 3	E2	146	170	16%	2.0	2.1	1%
	E3	83	165	90%	1.4	2.1	49%
Test 9	E4	75	167	124%			
	E1	71	142	99%	1.2	2.2	80%
Test 5	E2	118	144	23%	1.7	1.8	4%
	E3	90	167	94%	1.2	2.1	70%
Test 14	E2	148	169	14%	2.1	2.0	-2%
	E3	84	167	73%	1.4	2.1	46%
Test 15	E2	151	168	11%	2.2	2.0	-7%
	E3	87	166	76%	1.3	2.1	58%
Test 18	E2	153	167	9%	2.2	2.0	-9%

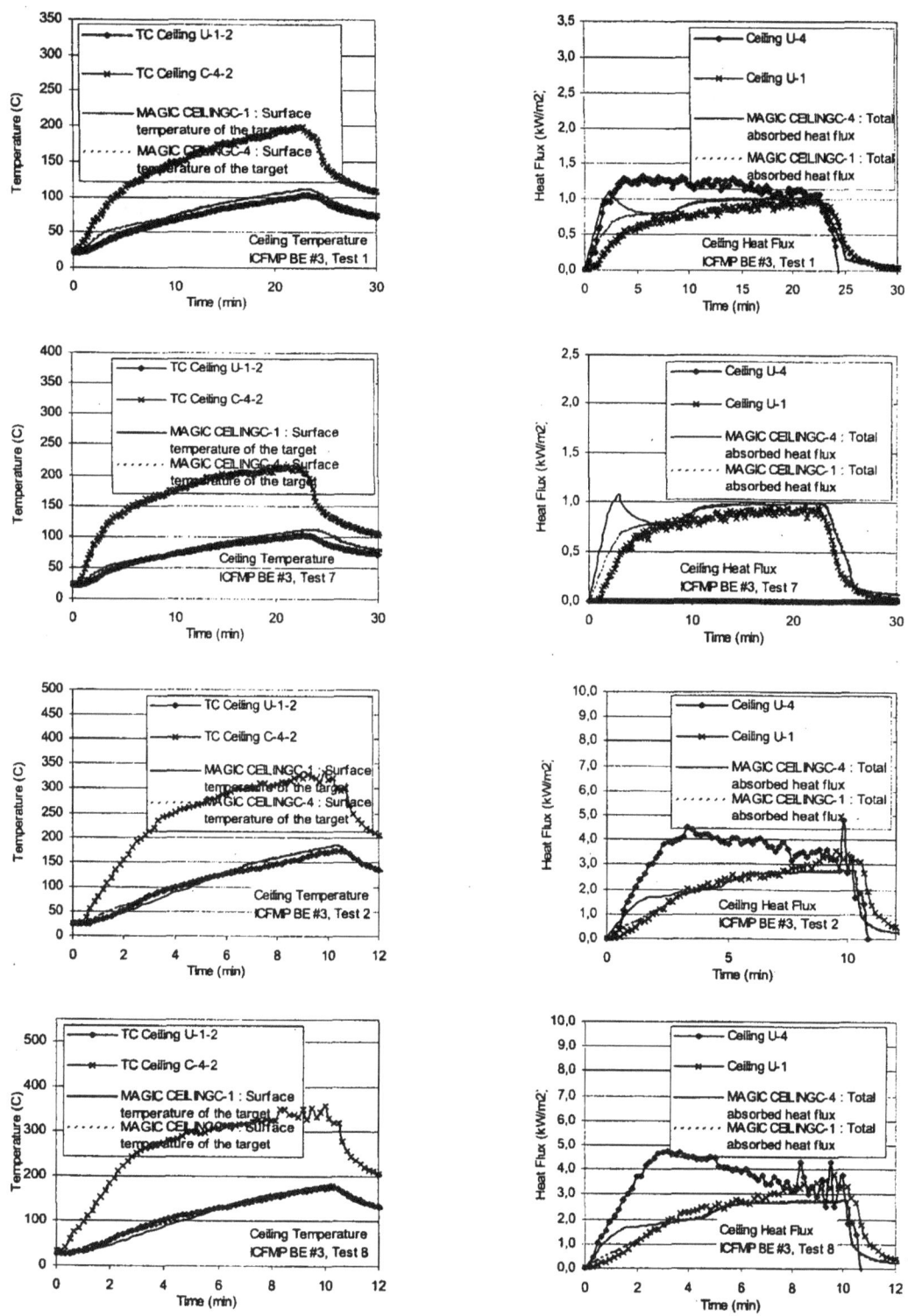

Figure A-78: Ceiling Heat Flux and Surface Temperature, ICFMP BE #3, Closed-Door Tests

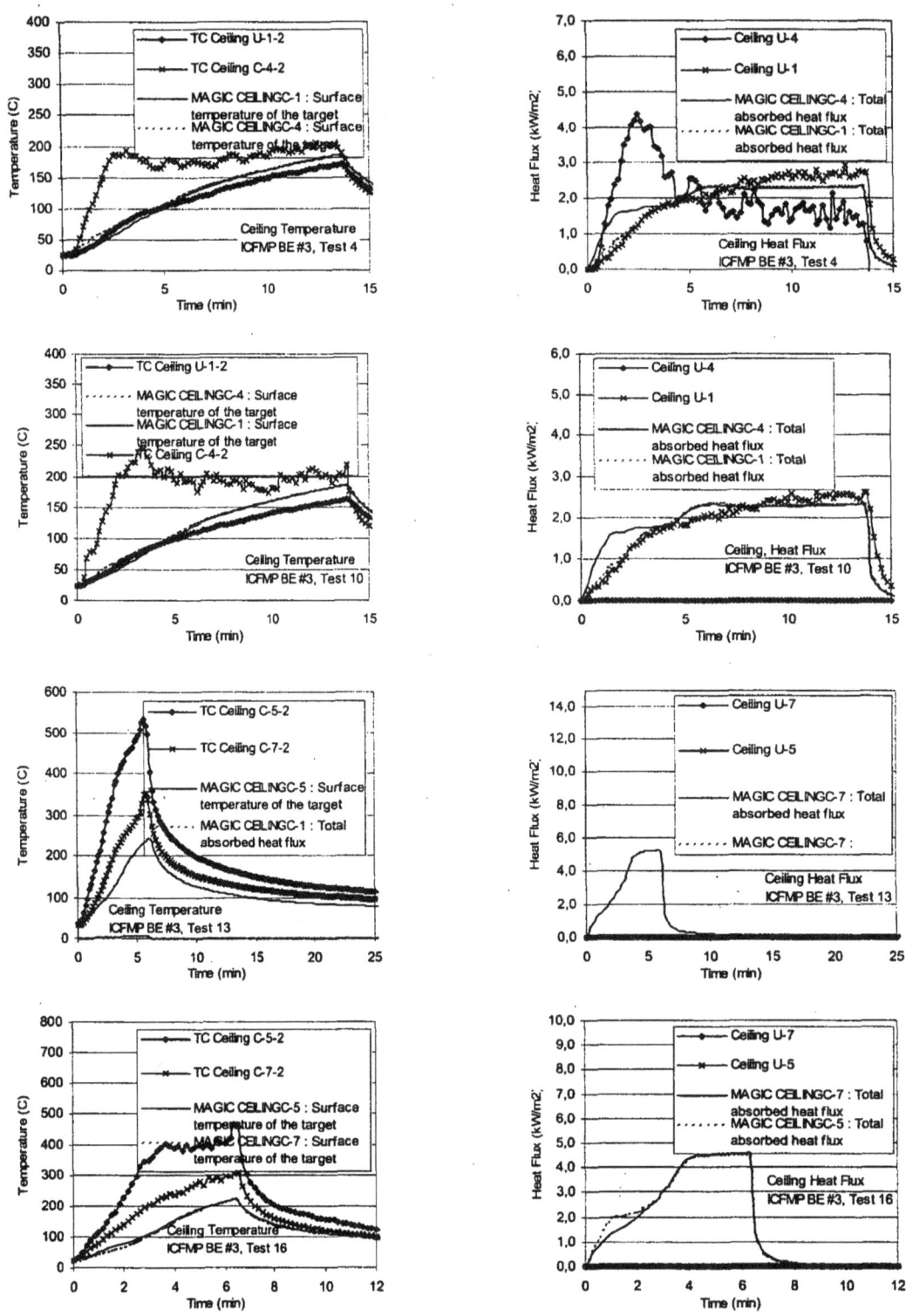

Figure A-79: Ceiling Heat Flux and Surface Temperature, ICFMP BE #3, Closed-Door Tests

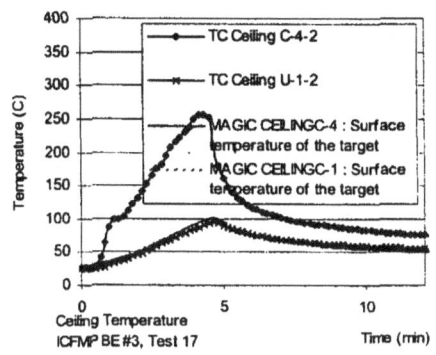

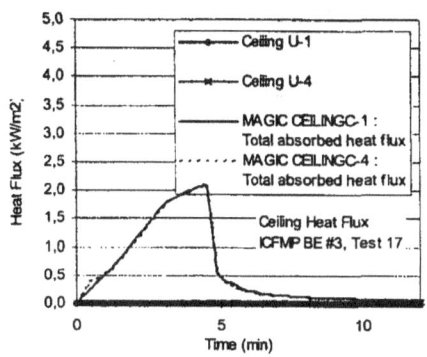

Open-door tests to Follow

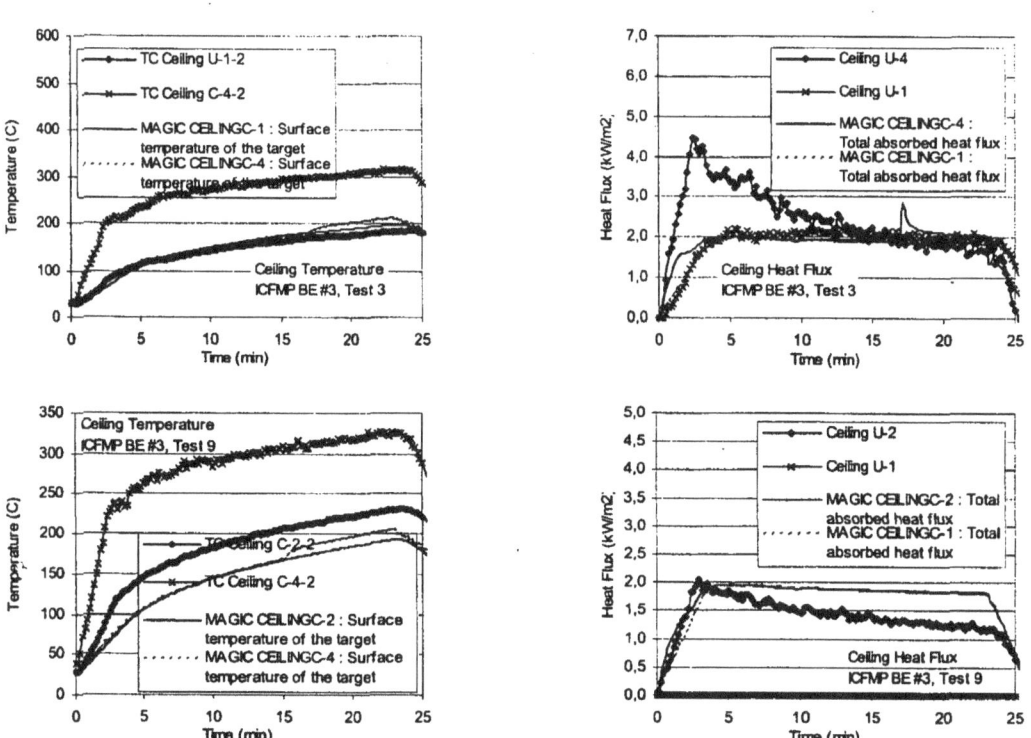

Figure A-80: Ceiling Heat Flux and Surface Temperature, ICFMP BE #3, Open-Door Tests

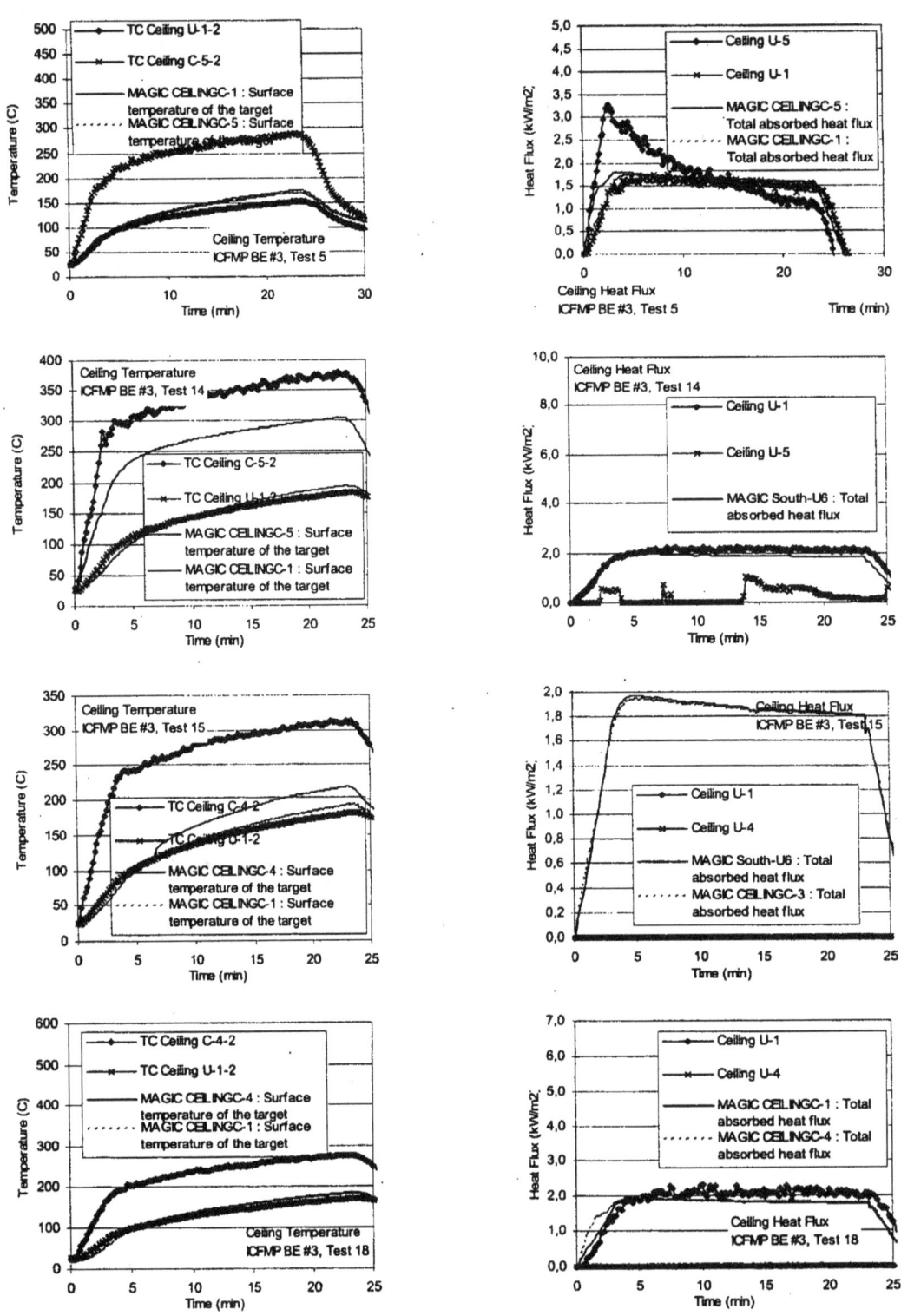

Figure A-81: Ceiling Heat Flux and Surface Temperature, ICFMP BE #3, Open-Door Tests

Table A-30: Relative Differences for Temperature and Total Heat Flux Corresponding to the Ceiling

Ceiling	Instrument	Ceiling Temperature			Ceiling Total Heat Flux		
		ΔE (°C)	ΔM (°C)	Relative Diff	ΔE (kW/m²)	ΔM (kW/m²)	Relative Diff
Test 1	C1	81	89	10%	1.0	1.0	3%
	C4	176	89	-49%	1.3	1.1	-17%
Test 7	C1	80	86	8%	1.0	1.0	-1%
	C4	191	87	-55%			
Test 2	C1	148	158	7%	3.5	2.8	-21%
	C4	308	159	-49%	4.8	2.8	-42%
Test 8	C1	148	157	6%	3.9	2.8	-28%
	C4	325	157	-52%	5.5	2.7	-50%
Test 4	C1	147	158	8%	2.9	2.4	-19%
	C4	180	159	-12%	4.9	2.3	-52%
Test 10	C1	138	158	14%	2.6	2.3	-10%
	C4	221	158	-29%			
Test 13	C7	35	209	501%			
	C5	500	210	-58%			
Test 16	C7	171	196	15%			
	C5	419	197	-53%			
Test 17	C1	69	71	4%			
	C4	230	71	-69%			
Test 3	C1	155	167	8%	2.2	2.0	-9%
	C4	287	181	-37%	4.5	2.8	-36%
Test 9	C2	46	166	260%	2.0	2.0	-2%
	C4	290	179	-38%	4.1	2.8	-32%
Test 5	C1	125	142	13%	2.0	1.8	-13%
	C5	166	146	-12%	4.7	1.8	-61%
Test 14	C1	158	166	5%	2.2	2.0	-11%
	C5	248	277	12%			
Test 15	C1	157	166	6%			
	C4	287	192	-33%			
Test 18	C1	145	165	14%	2.3	2.0	-16%
	C4	250	167	-33%			

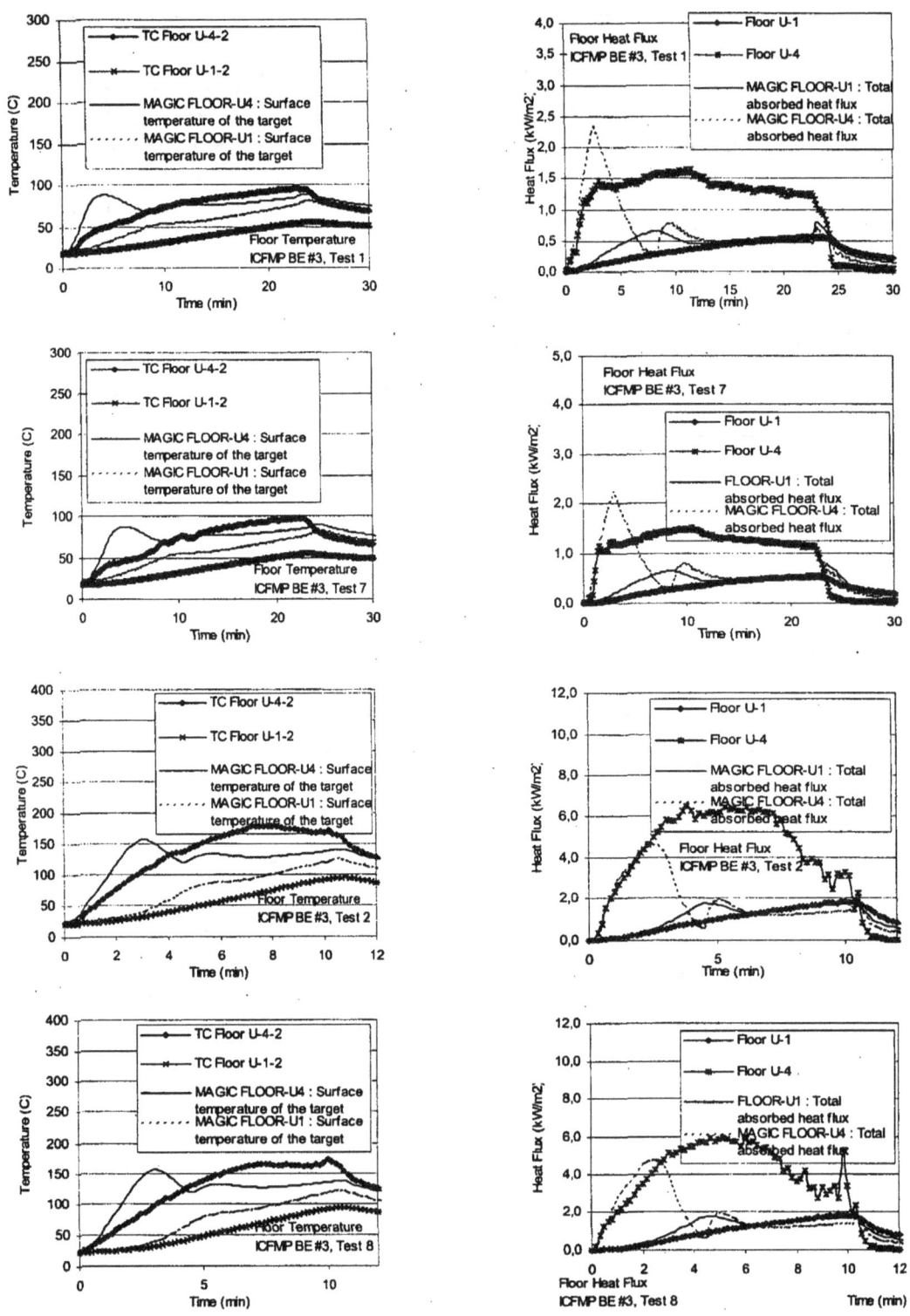

Figure A-82: Floor Heat Flux and Surface Temperature, ICFMP BE #3, Closed-Door Tests

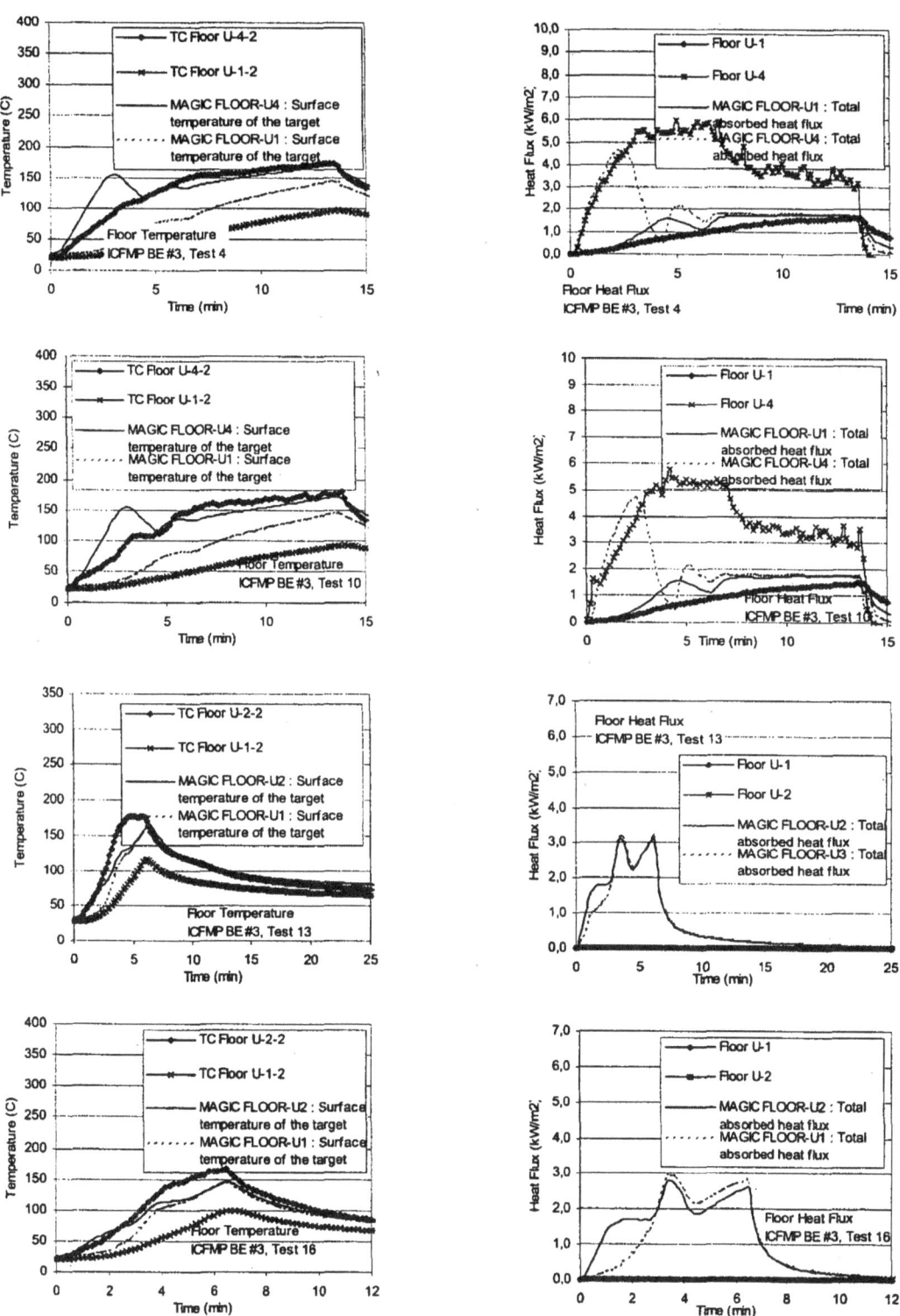

Figure A-83: Floor Heat Flux and Surface Temperature, ICFMP BE #3, Closed-Door Tests

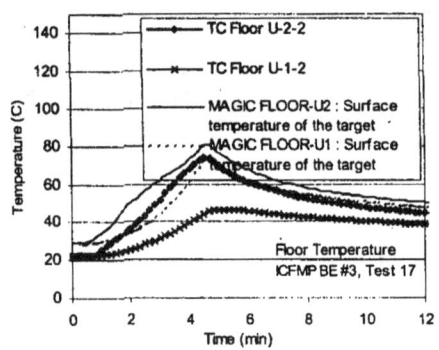
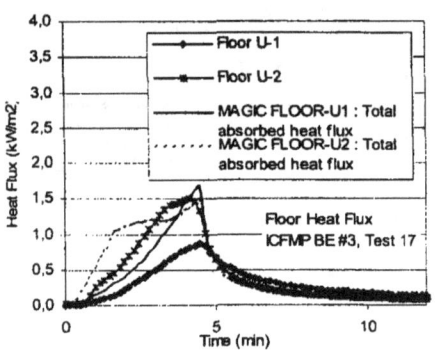

Open-door tests to Follow

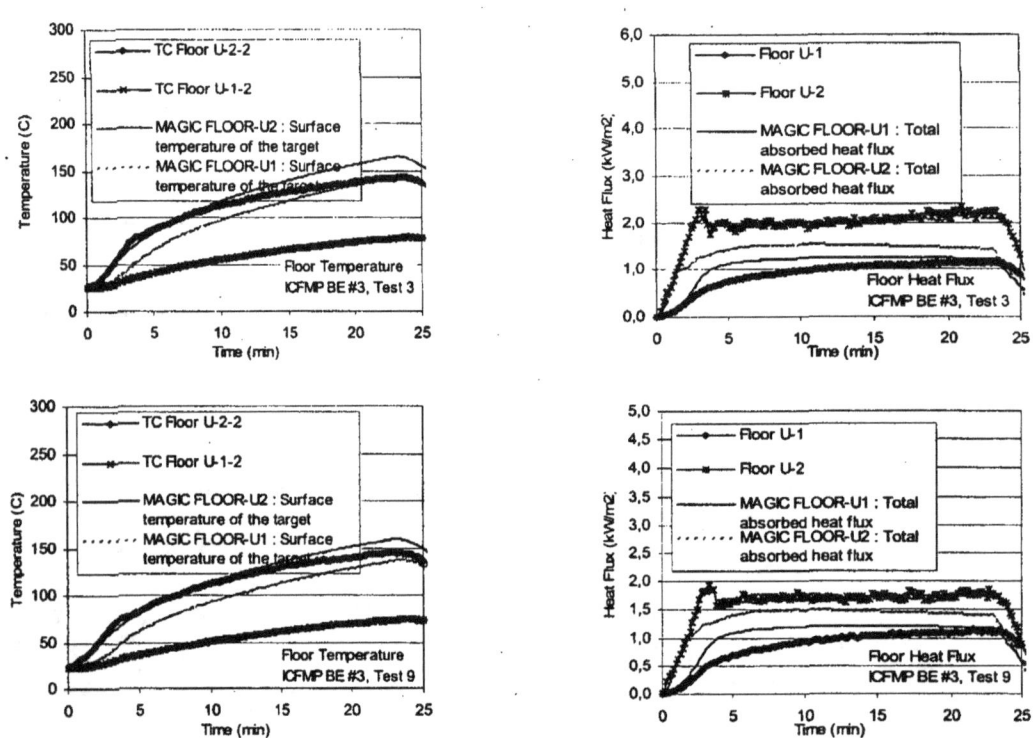

Figure A-84: Floor Heat Flux and Surface Temperature, ICFMP BE #3, Open-Door Tests

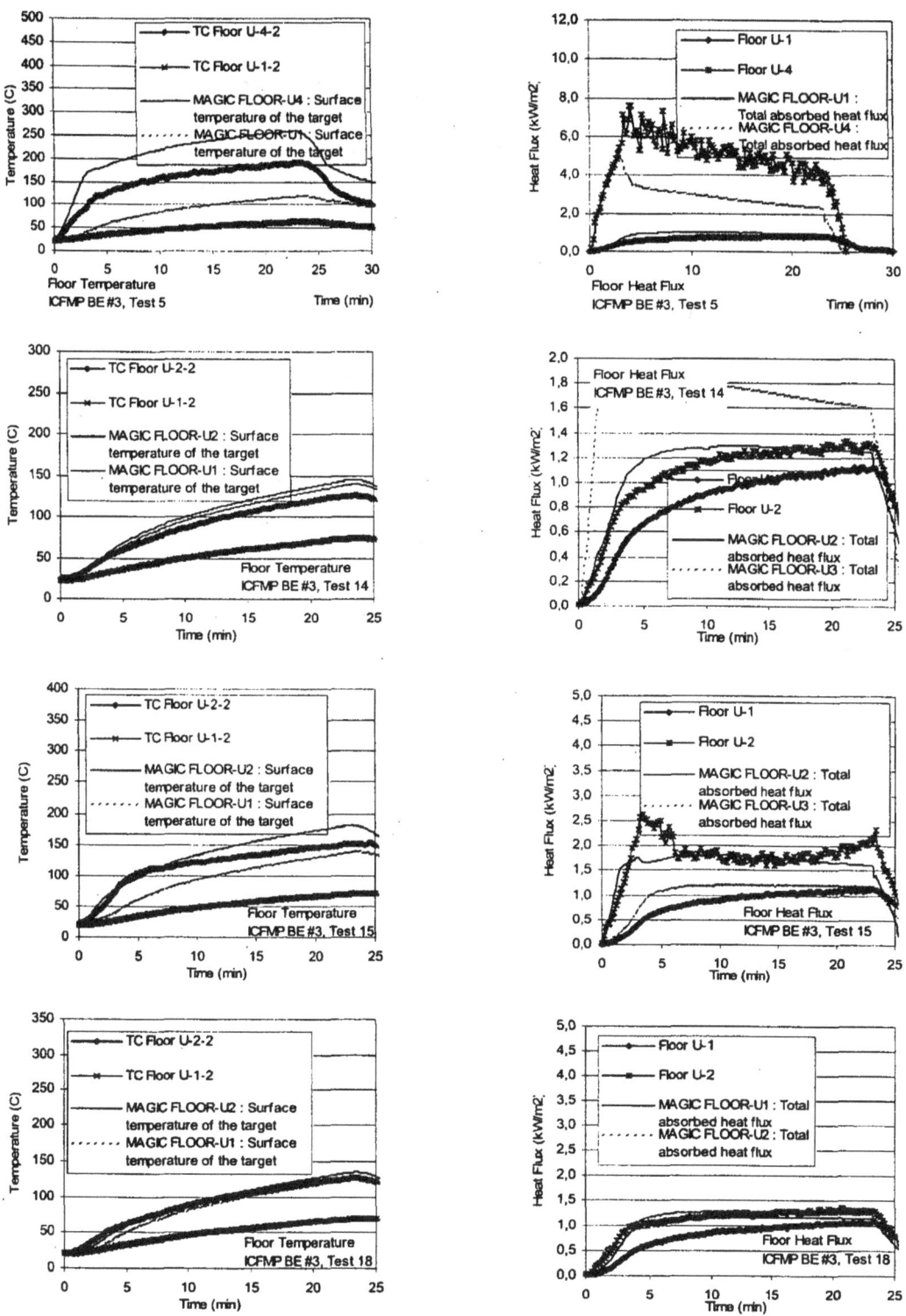

Figure A-85: Floor Heat Flux and Surface Temperature, ICFMP BE #3, Open-Door Tests

Table A-31: Relative Differences for Temperature and Total Heat Flux Corresponding to the Floor

Floor	Instrument	Floor Temperature			Floor Total Heat Flux		
		ΔE (°C)	ΔM (°C)	Relative Diff	ΔE (kW/m^2)	ΔM (kW/m^2)	Relative Diff
	F1	38	53	40%	0.6	0.7	18%
Test 1	F4	77	66	-15%	1.6	2.3	43%
	F1	36	52	43%	0.6	0.6	9%
Test 7	F4	78	63	-19%	1.7	2.3	34%
	F1	74	99	34%	1.8	1.9	6%
Test 2	F4	156	131	-16%	6.4	4.7	-27%
	F1	71	98	37%	1.9	1.9	2%
Test 8	F4	148	132	-11%	6.2	4.8	-23%
	F1	76	118	55%	1.6	1.7	7%
Test 4	F4	152	144	-5%	5.9	4.7	-22%
	F1	71	120	68%	1.5	1.7	17%
Test 10	F4	158	145	-8%	5.7	4.7	-17%
	F1	89	132	48%			
Test 13	F2	73	130	77%			
	F1	80	119	49%			
Test 16	F2	206	118	-43%			
	F1	24	44	80%	0.9	1.7	96%
Test 17	F2	117	51	-56%	1.5	1.5	1%
	F1	54	112	110%	1.2	1.2	8%
Test 3	F2	186	135	-27%	2.3	1.5	-34%
	F1	53	110	106%	1.2	1.2	5%
Test 9	F2	94	132	41%	1.9	1.5	-21%
	F1	42	91	118%	0.9	1.0	16%
Test 5	F4	171	232	36%	10.0	4.9	-52%
	F1	52	111	113%	1.1	1.2	9%
Test 14	F2	47	117	149%	1.3	1.3	-3%
	F1	52	112	113%	1.2	1.2	6%
Test 15	F2	140	155	10%	7.5	1.8	-76%
	F1	50	108	118%	1.1	1.2	10%
Test 18	F2	55	117	115%	1.3	1.3	-4%

A.9.2 ICFMP BE #4

Three thermocouples were mounted on the back wall of the compartment. Because the fire leaned toward the back wall, the temperatures measured by the thermocouples are considerably higher than for most of the other wall surface points considered in this report.

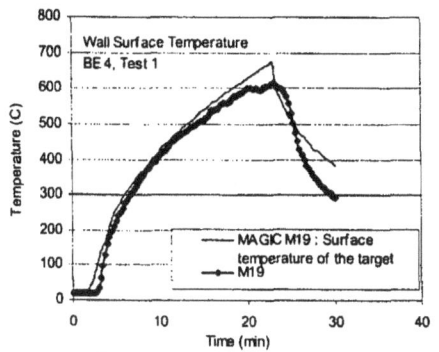 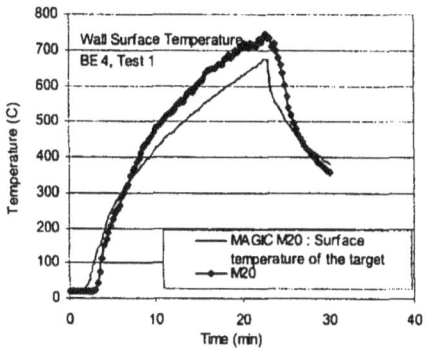

Figure A-86: Back Wall Surface Temperatures, ICFMP BE #4

Table A-32: Relative Differences for Wall Temperature

	Instrument	ΔE (°C)	ΔM (°C)	Relative Diff
ICFMP 4-1	M19	596	656	10%
	M20	724	656	-9%

A.9.3 ICFMP BE #5

Wall surface temperatures were measured in two locations during the BE #5 test series. The thermocouples labeled TW 1-x (Wall Chain 1) were against the back wall; those labeled TW 2-x (Wall Chain 2) were behind the vertical cable tray. Seven thermocouples were in each chain, spaced 80 cm (31.5 inches) apart. In Figure A-87, the lowest (1), middle (4), and highest (7) locations are used for comparison.

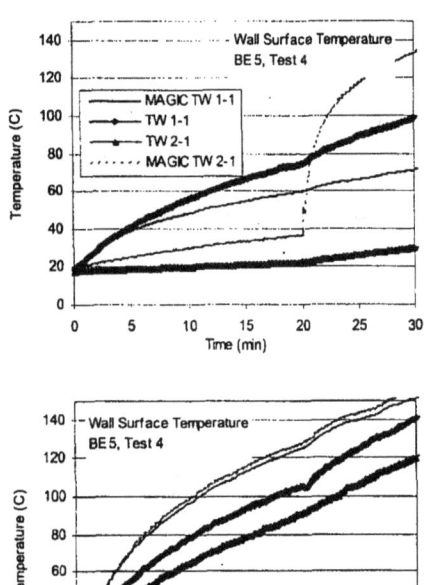
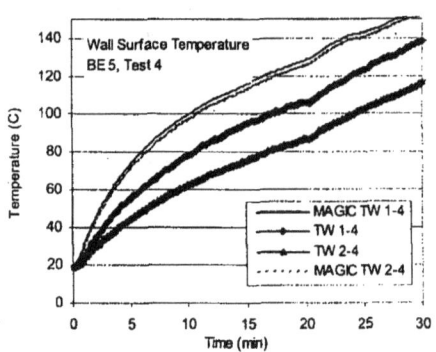

Figure A-87: Back and Side Wall Surface Temperatures, ICFMP BE #5, Test 4

Table A-33: Relative Differences for Wall Temperature

	Instrument	ΔE (°C)	ΔM (°C)	Relative Diff
ICFMP 5-4	TW 1-1	79	51	-35%
	TW 2-1	12	113	868%
	TW 1-4	118	134	13%
	TW 2-4	96	132	38%
	TW 1-7	121	131	9%
	TW 2-7	100	135	36%

B
MAGIC INPUT FILES

This appendix is reserved for the MAGIC input files used for the simulations in this V&V study. These files are only available electronically because of their size and formatting.

NRC FORM 335 (9-2004) NRCMD 3.7	U.S. NUCLEAR REGULATORY COMMISSION	1. REPORT NUMBER (Assigned by NRC, Add Vol., Supp., Rev., and Addendum Numbers, if any.)
BIBLIOGRAPHIC DATA SHEET *(See instructions on the reverse)*		NUREG-1824

2. TITLE AND SUBTITLE

Verification and Validation of Selected Fire Models for Nuclear Power Plant Applications
Volume 6: MAGIC

3. DATE REPORT PUBLISHED

MONTH	YEAR
May	2007

4. FIN OR GRANT NUMBER

5. AUTHOR(S)

F. Joglar (EPRI/SAIC), B. Gautier (EdF), L. Gay (EdF), J. Texeraud (EdF)

6. TYPE OF REPORT

Technical

7. PERIOD COVERED *(Inclusive Dates)*

8. PERFORMING ORGANIZATION - NAME AND ADDRESS *(If NRC, provide Division, Office or Region, U.S. Nuclear Regulatory Commission, and mailing address; if contractor, provide name and mailing address.)*

U.S. Nuclear Regulatory Commission, Office of Regulatory Research (RES), Washington, DC 20555-0001

Electric Power Research Institute (EPRI), 3412 Hillview Avenue, Palo Alto, CA 94303

Science Applications International Corp. (SAIC), 4920 El Camino Real, Los Altos, CA 94022

National Institute of Standards and Technology (NIST/BFRL), 100 Bureau Drive, Stop 8600, Gaithersburg, MD 20899-8600

9. SPONSORING ORGANIZATION - NAME AND ADDRESS *(If NRC, type "Same as above"; if contractor, provide NRC Division, Office or Region, U.S. Nuclear Regulatory Commission, and mailing address.)*

U.S. Nuclear Regulatory Commission, Office of Regulatory Research (RES), Washington, DC 20555-0001

Electric Power Research Institute (EPRI), 3412 Hillview Avenue, Palo Alto, CA 94303

10. SUPPLEMENTARY NOTES

11. ABSTRACT *(200 words or less)*

There is a movement to introduce risk-informed and performance-base analyses into fire protection engineering practice, both domestically and worldwide. The move towards risk-informed decision-making in nuclear power regulation was directed by the U.S. Nuclear Regulatory Commission.

One key tool needed to support risk-informed, performance-based fire protection is the availability of verified and validated fire models that can accurately predict the consequences of fires. Section 2.4.1.2. of NFPA 805, Performance-Base Standard for Fire Protection for Light-Water Reactor Electric Generating Plants, 2001 Edition requires that only fire models acceptable to the Authority Having Jurisdiction (AHJ) shall be used in fire modeling calculations. Futhermore, Sections 2.4.1.2.2. and 2.4.1.2.3. of NFPA 805 state that fire models shall be applied within the limitations of the given model, and shall be verified and validated. This report is the first effort to document the verification and validation (V&V) of five models that are commomly used in nuclear power plant applications. The project was performed in accordance with the guidelines that the American Society for Testing and Materials (ASTM) set forth in ASTM E 1355, Standard Guide for Evaluating the Predictive capability of Deterministic Fire Models. The results of this V&V are reported in the form of color codes describing the accuracies for the model predictions.

12. KEY WORDS/DESCRIPTORS *(List words or phrases that will assist researchers in locating the report.)*

fire, fire modeling, verification, validation, performance-based, risk-informed, firehazards analyses, V&V, FHA, CFAST, FDS, MAGIC, FIVE, FDTs

13. AVAILABILITY STATEMENT

unlimited

14. SECURITY CLASSIFICATION

(This Page)

unclassified

(This Report)

unclassified

15. NUMBER OF PAGES

16. PRICE

PRINTED ON RECYCLED PAPER

www.ingramcontent.com/pod-product-compliance
Lightning Source LLC
Chambersburg PA
CBHW081441170526
45166CB00008B/2274